PRAISE FOR *THE WAY WE EAT NOW*

"Nobody else writing about our global food landscape is as fearless, rigorous, compassionate, or readable as Bee Wilson. Thank God we have her. Why we eat what we eat—and what has shaped our global food landscape—is one of the most vital questions of our time."

—Diana Henry, author of *Simple: Effortless Food, Big Flavors*

"Bee Wilson's fascinating book is a guide to the future of food and how we eat and will be eating. Meticulously researched and yet written brilliantly for the layman, her book will be often consulted on my bookshelf!"

—Ken Hom OBE, chef, author, and TV presenter

THE WAY
WE EAT NOW

Also by Bee Wilson:

Consider the Fork: A History of How We Cook and Eat
First Bite: How We Learn to Eat

THE WAY
WE EAT NOW

*How the Food Revolution Has Transformed
Our Lives, Our Bodies, and Our World*

BEE WILSON

BASIC BOOKS
New York

Cover design by Chin-Yee Lai
Cover images © Annabelle Breakey / Getty Images; © Shana Novak / Offset.com
Cover copyright © 2019 Hachette Book Group, Inc.

Basic Books
Hachette Book Group
1290 Avenue of the Americas, New York, NY 10104
www.basicbooks.com

Printed in the United States of America
First Edition: May 2019

Published by Basic Books, an imprint of Perseus Books, LLC, a subsidiary of Hachette Book Group, Inc. The Basic Books name and logo is a trademark of the Hachette Book Group.

The Hachette Speakers Bureau provides a wide range of authors for speaking events. To find out more, go to www.hachettespeakersbureau.com or call (866) 376-6591.

The publisher is not responsible for websites (or their content) that are not owned by the publisher.

Print book interior design by Amnet Systems.

Library of Congress Cataloging-in-Publication Data

Names: Wilson, Bee, author.
Title: The way we eat now : how the food revolution has transformed our
    lives, our bodies, and our world / Bee Wilson.
Description: First edition: May 2019. | New York : Basic Books, 2019. |
    Includes bibliographical references and index.
Identifiers: LCCN 2018047349 | ISBN 9780465093977 (hardcover)
Subjects: LCSH: Gastronomy. | Food habits. | Food. | Diet.
Classification: LCC TX631 .W5484 2019 | DDC 641.01/3—dc23
LC record available at https://lccn.loc.gov/2018047349

ISBNs: 978-0-465-09397-7 (hardcover), 978-0-465-09398-4 (ebook)

LSC-C

10 9 8 7 6 5 4 3 2 1

For Leo

Ever the eaters and drinkers, ever the upward and downward sun, ever the air and the ceaseless tides
—Walt Whitman, "Song of Myself"

CONTENTS

INTRODUCTION:
THE GATHERERS AND
THE HUNTED

PICK A BUNCH OF GREEN GRAPES, WASH IT, AND PUT ONE in your mouth. Feel the grape with your tongue; observe how cold and refreshing it is: the crisp flesh and the jellylike interior with its mild, sweet flavor.

Eating grapes can feel like an old pleasure, untouched by change. The ancient Greeks and Romans loved to eat grapes, as well as to drink them in the form of wine. *The Odyssey* speaks of "a ripe and luscious vine, hung thick with grapes." As you pull the next delicious grape from its stalk, you could easily be plucking it from a Dutch still life of the seventeenth century, where grapes are tumbled on a metal platter with oysters and half-peeled lemons.

But look closer at this bunch of green grapes, cold from the fridge, and you see that this fruit is not unchanged after all. Like so many other foods, grapes have become a product of engineering designed to please modern eaters. First of all, there are almost certainly no grape seeds for you to either chew or spit out (unless you are in certain places, such as Spain or China, where seeded grapes are still tolerated). Strains of seedless grapes have been cultivated for centuries, but it is only in the past two decades that seedless has become the norm, to spare us the dreadful inconvenience of seeds.

Here's another strange new thing about grapes: the mainstream ones in supermarkets, such as Thompson Seedless and Crimson Flame, are always sweet—not bitter, acidic, or foxy like a Concord grape, or excitingly aromatic like one of the Muscat varieties of Italy, but just plain sweet, like sugar. On biting into a grape, the ancients did not know if it would be ripe or sour. The same was true, in my experience, as late as the 1990s. It was like grape roulette: a truly sweet one was rare and therefore special.

These days, the sweetness of grapes is a sure bet because in common with other modern fruits, such as red grapefruit and Pink Lady apples, our grapes have been carefully bred and ripened to appeal to consumers reared on sugary foods. Fruit bred for sweetness does not necessarily have to be less nutritious, but modern de-bittered fruits tend to contain fewer of the phytonutrients that give fruits and vegetables many of their protective health benefits. Most of the phytonutrients in green grapes were in the seeds. A modern red or purple seedless grape will still be rich in phenolics—nutrients that reduce the risk of certain cancers—from the pigments in its skin. But green seedless grapes contain few of these phytonutrients at all. Such fruit still gives us energy but not necessarily the health benefits we would expect.[1]

The very fact that you are nibbling seedless grapes so casually is also new. I am old enough to remember a time when grapes—unless you were living in a grape-producing country—were a special and expensive treat. But now, millions of people on average incomes can afford to behave like the reclining Roman emperor of TV cliché, popping grapes into our mouths one by one. Globally, we both produce and consume *twice* as many grapes as we did in the year 2000. Grapes are an edible sign of rising prosperity because fruit is one of the first little extras that people spend money on when they start to have disposable income. The year-round availability of grapes also speaks to huge changes in global agriculture. Fifty years ago, table grapes were a seasonal fruit, grown in just a few countries and eaten only at certain times of year. Today, they are cultivated globally and never out of season.[2]

Almost everything about grapes has changed, and fast. And yet grapes are the least of our worries when it comes to food: just one tiny element in a much larger series of kaleidoscopic transformations in how and what we eat that have happened in recent years. These changes are written on the land, on our bodies, and on our plates (insofar as we even eat off plates anymore).

FOR MOST PEOPLE ACROSS THE WORLD, LIFE IS GETTING better but diets are getting worse. This is the bittersweet dilemma of eating in our times. Unhealthy food, eaten in a hurry, seems to be the price we pay for living in liberated modern societies. Even grapes—so sweet, so convenient, so ubiquitous—are symptoms of a food supply that is out of control. Millions of us enjoy lives that are freer and more comfortable than those our grandparents lived, a freedom underpinned by the amazing decline in global hunger. You can measure this life improvement in many ways, whether by the growth of literacy and smartphone ownership, the spread of labor-saving devices such as dishwashers, or the rising number of countries where gay couples have the right to marry. Yet our free and comfortable lifestyles are undermined by the fact that our food is killing us, not through its lack but through its abundance—a hollow kind of abundance.*

---

* Some would argue that the fact that we tend to die of chronic noncommunicable diseases rather than acute infectious ones is a sign of how lucky we are. All mortals have to die of something, and it is progress that compared to twenty-five years ago, so many people are living longer lives and therefore dying from chronic conditions such as heart disease and cancer rather than dying in childhood from acute hunger or unsafe drinking water. What is not progress

What we eat now is a greater cause of disease and death in the world than either tobacco or alcohol. In 2015, around 7 million people died from tobacco smoke and 3.3 million from causes related to alcohol, but 12 million deaths could be attributed to "dietary risks" such as those that arise from diets low in vegetables, nuts, whole grains, and seafood or diets high in salt (mostly from processed food) and sugary drinks. This is paradoxical and sad, because good food—good in every sense, from flavor to nutrition—used to be the test by which we judged the quality of life. A good life without good food should be a logical impossibility.[3]

Where humans used to live in fear of plague or tuberculosis, now the leading cause of mortality worldwide is diet.[4] Most of our problems with eating come down to the fact that we have not yet adapted to the new realities of plenty, either biologically or psychologically. Many of the old ways of thinking about diet no longer apply, but it isn't clear yet what it would mean to adapt our appetites and routines to the new rhythms of life. We take our cues about what to eat from the world around us, which becomes a problem when our food supply starts to send us crazy signals about what is normal. "Everything in moderation" doesn't quite cut it in a world where the "everything" for sale in the average supermarket has become so sugary and so immoderate. In today's world, it can be hard to know how to eat for the best. Some binge; some restrict. Some put their faith in expensive "superfoods" that promise to do things for the human body that mere food cannot. Others—this is how far things have gone—have lost faith in solid food altogether, choosing instead to drink one of the new meal-replacement beverages—curious beige liquids that have become an aspirational form of nutrition.

---

is that so much death and disease in the world should be caused by poor diet, which is a preventable cause of human suffering. Across the world, NCDs cause 80 percent of all DALYS—disability-adjusted life years, a figure that refers to the years of life blighted by premature illness.

To our grandparents, it would not have seemed credible for any hungry human being to think that *not eating* was a better option than eating. But our grandparents never had to live and eat in a bewildering and complex food culture such as ours.

At no point in history have edible items been so easy to obtain, and in many ways, this is a glorious thing. Humans have always gone out and gathered food, but never before has it been so simple for us to gather anything we want, whenever and wherever we want it, from sachets of squid ink to strawberries in winter. We can get sushi in Buenos Aires, sandwiches in Tokyo, and Italian food everywhere. Not so long ago, to eat genuine Neapolitan pizza, a swollen-edged disk of dough cooked in a blistering oven, you had to go to Naples. Now, you can find Neapolitan pizza—the real deal, cooked in a suitably hot oven using the right dough—as far afield as Seoul and Dubai. Thanks to the new home delivery apps such as Deliveroo and Seamless, we can have food from almost any cuisine on our doorstep in minutes.

The gatherers of the world never had it so good. In our hunter-gatherer past, if you wanted a taste of something sweeter than fruit, you called a group of brave comrades together and went on a long, perilous expedition, scrambling up rocks, hunting in crevices for wild honey. Often, the honey hunters came back empty-handed. Now, if you fancy a taste of something sweet, you head to the nearest shop with a little loose change. You do not come back empty-handed.

THE FLIPSIDE OF FOOD BEING SO EASY TO GET IS THAT IT is also hard to escape.

We are the first generation to be hunted by what we eat. Since the birth of farming ten thousand years ago, most humans haven't been hunters, but never before have we been so insistently pursued by our own food supply. The calories hunt us down even when we are not looking for them. They tempt us at the supermarket checkout and on the coffee shop counter. They sing to us in adverts when we switch on the TV. They track us down on social media with amusing videos that make us want even more. They sneak into our mouths as free samples. They console us for our pains, only to become the cause of fresh sorrows. They trick us by hiding in "healthy snacks" for our children that are just as high in sugar as the "unhealthy" alternatives.

Talking about what has gone wrong with modern eating is delicate, because food is a touchy subject. No one likes to feel judged about their food choices, which is one of the reasons why so many healthy-eating initiatives fail. The foods that are destroying our health are often the ones to which we feel the deepest emotional connection. They are the stuff of childhood memories. Some say we should never speak of "junk food" because it is a pejorative term to use about someone else's pleasures. But when poor diets become the single greatest cause of death in the world, I think we are allowed to be pejorative—not about our fellow eaters, but about the products that are making people so unwell.[5]

The rise of obesity and diet-related disease around the world has happened hand in hand with the marketing of fast food and sugary sodas, of processed meats and branded snack foods. As things stand, our culture is far too critical of the individuals who eat junk foods and not critical enough about the corporations that profit from selling them. We spend a lot of time discussing unhealthy foods in terms of individual guilt and willpower and not enough looking at the morality of big food companies that have targeted some of the poorest consumers in the world with products that will make them sick, or the governments that allowed them to do so. A survey of more than three hundred international policy makers found that 90 percent of them still believed that personal motivation—a.k.a. willpower—was a very strong cause of obesity.[6] This is nonsensical.

It makes no sense to presume that there has been a sudden collapse in willpower across all ages and ethnic groups and each sex since the 1960s. What has changed most since the sixties is not our collective willpower but the marketing and availability of energy-dense, nutrient-poor foods. Some of these changes are happening so rapidly it's almost impossible to keep track. Sales of fast food grew by 30 percent worldwide from 2011 to 2016, and sales of packaged food grew by 25 percent. Somewhere in the world, a new branch of Domino's pizza opened every seven hours in 2016.[7]

Compared to even five years ago, the quantities in which sweets are marketed are obscene. Oversized chocolate bars are nothing new, but I was stunned recently at my nearest supermarket to see Snickers chocolate being sold not by the bar, not even by the supersized bar, but by the *meter*, consisting of ten bars joined together: 2,340 calories of chocolate, on special offer for a couple of dollars. If that is not an incitement to overeat, I don't know what is.

Encouraging us to buy more food than we intend or need is a large part of the business strategy of all the major food companies. Until the mid-1990s, Hank Cardello advised some of the biggest food producers in the world. Cardello reveals that the mantra of the packaged food companies was that "you could make Americans eat just about anything, so long as you sold it right." When the Western appetite for packaged foods finally started to reach saturation point, the industry moved on to new markets overseas. In developing and middle-income countries, branded food now hunts people down even in the privacy of their own homes. Through direct sales, multinational food companies are aggressively targeting low-income customers in some of the world's remotest villages.[8]

It isn't that food executives are evil people who actively set out to make their customers obese. But as Cardello has explained, for too long, the well-being of consumers simply didn't figure in the calculations of the big food and beverage companies that he worked with. "All we thought about was market expansion and our own bottom line."[9] Food and beverage manufacturers explicitly talk among themselves of "heavy users" as representing their key clientele: when it comes to sugary drinks and sweets, 80 percent of the product is bought by just 20 percent of the customers.[10] "Heavy user" is industry speak for people suffering from binge-eating disorder.

Yet junk food is far from the only cause of obesity, whose roots are complex and multifaceted. Across the board, across all social classes, most of us eat and drink more than our grandparents did, whether we are cooking a leisurely dinner at home from fresh ingredients or grabbing a quick takeaway from a fast-food chain. Plates are bigger than they were fifty years ago, our idea of a portion is inflated, and wine glasses are vast. It's become normal to punctuate the day with snacks and to quench our thirst with a series of calorific liquids, from green juice and detox shots to craft sodas. You can gain weight eating expensive organic artisanal apple tarts and huge mugs of milky coffee just as easily as you can eating cheap fried chicken and Coke. As the example of grapes shows, we don't just eat more burgers and fries than our grandparents. We also eat more fruit and more granola bars, more avocado toast and more frozen yogurt, more salad dressing, and many, many more "guilt-free" kale chips.[11]

Almost every country in the world has experienced radical changes to its patterns of eating over the past five, ten, and fifty years. Taken together, these changes represent a food revolution in which none of the old certainties about eating look so certain anymore. For a long time, nutritionists have held up the Mediterranean diet as a healthy model for people in all countries to follow. But recent reports from the World Health Organization suggest that even in Spain, Italy, and Crete, most children no longer eat anything like a Mediterranean diet rich in olive oil and fish and tomatoes.[12] These Mediterranean children, who are, as of 2017, among the most overweight in Europe, now drink sugary colas and eat packaged snack foods; they have lost the taste for fish and olive oil. On every continent, there has been a common set of changes from savory foods to sweet ones, from meals to snacks, from small independent food shops to giant supermarkets, from dinners cooked at home to meals eaten out or as takeout.

THESE CAN BE SCARY AND CONFUSING TIMES IN WHICH to eat, made still scarier by the fact that there are so many "experts" out there selling us fear of food and fad cures. Times of transition have always been a gift to confidence tricksters.[13] When everything seems to be changing and we can no longer rely on the truths of the past, we become vulnerable to hucksters. Some diet gurus tell us to beware all grains; others tell us that we should fear supposedly "acid-producing" foods ranging from dairy to meat and coffee. These new diets are perhaps best seen as a dysfunctional response to a still more dysfunctional food supply: a false promise of purity in a toxic world. Meanwhile, eating disorders are on the rise across the world, among men as well as women.

Happiness at the table entails making your peace with food, and so it's a worrying development that eating now is so often treated as an all-or-nothing game. Food has never been so angrily polarized into virtues and vices, elixirs and poisons. On a single street in a single town, there will be some people eating giant burgers toppling with many layers of meat and sauces and others eating supposedly perfect meals of kale and seaweed with fermented kombucha to drink. There are gurus telling us to avoid gluten "just in case" and others teaching us to be frightened of cheese. I worry that in many cases, our pursuit of the perfect meal has become the enemy of the good-enough meal. While we fixate on this or that wonder ingredient, the thing that seems to be in short supply now is the everyday, unglamorous home-cooked dinner.

Part of the problem is that we have lost our trust in our own senses to tell us what to eat. We wouldn't be such easy prey for extreme diets if we could recognize food when it was right underneath our noses. Humans seem to have become—both collectively and individually—very poor at identifying food when we see it, partly because so much of what our culture offers up for us to taste is so heavily packaged and disguised.

If we have lost knowledge about we are actually eating, we have also lost the old norms regarding how to eat it. Sometimes this looks like freedom; sometimes, like chaos. In 1958, survey data suggests, nearly three-quarters of British adults drank hot tea with the evening meal, because this was the expected way to behave. Now, such shared expectations about food have largely vanished. Who can say for certain when "lunchtime" really is anymore? This generation has lived through revolutionary changes not just in what we eat but also how we eat it. Our appetites used to be held in place by a series of invisible threads, rituals that told us how to behave when we held a knife and fork. Now, the rituals are mostly gone, and so are the knives and forks.[14]

The nutrient content of our meals is one thing that has radically changed; the psychology of eating is another. Much of our eating takes place in a new chaotic atmosphere in which we no longer have many rules to fall back on. The problem is partly that cooking at home from raw ingredients is no longer the unquestioned daily routine it once was. One of the functions of traditional cuisines was to create a common understanding of what ingredients could and couldn't be combined. Sometimes, these rules could feel restrictive and annoying, such as the Italian insistence that fish and cheese can never enhance each other (tell that to the person who has just enjoyed a delicious fish pie with cheddar cheese on top). But at least these culinary rules gave a sense of structure to our eating, whether you obeyed them or not. Now, many of us are eating with no structure to guide us, with the day passing in a blur of bizarre snacks. When I interviewed a product developer for a major UK supermarket in 2017, she said that the main way that British eating behavior had changed over the past decade was that people had become so erratic and hard to categorize. In a single basket of food, shoppers oscillate wildly between vegan health foods such as oat milk and meat-heavy "dude food" such as pizzas topped with pulled pork.

On an early evening train journey recently, I looked up at my fellow travelers and noticed, first, that almost everyone was eating or drinking something and, second, that they were all doing so in ways that might once have been considered deeply eccentric. One man had both a cappuccino and a can of fizzy drink from which he was taking alternate sips. A woman with headphones on was nibbling an apricot tart, produced from a cardboard patisserie box. She followed it with a high-protein snack of two hard-boiled eggs and some raw spinach. Sitting across from her was a man carrying a worn leather briefcase. He reached inside the case and produced a bottle of strawberry milk-shake and a half-finished packet of chocolate-caramel sweets.

Like other modern eaters, these travelers were improvising their own food rules as they went along. The most surprising thing about this scene—which took place between Birmingham and London—is that it could have happened on a train between cities almost any-where. As I first embarked on this book, my plan was to explore how people eat in very different ways around the world. But as I met peo-ple from different countries, I kept being struck that the things they told me about modern eating were, to a bizarre extent, the same. This is another paradox of our times. Most people can afford to eat a more varied diet than in the past, but our varied diets are varied in the same way. From Mumbai to Cape Town, from Milan to Nanjing, people told me they felt they had lived through huge changes in the way they ate compared to their parents and certainly compared to their grandparents. They spoke of the erosion of traditional home cooking and the rise of McDonald's and of eating in front of screens. They also spoke of the backlash against ultra-processed food and the way that certain "healthy" foods (notably quinoa) had become a fetish of late. They spoke of weight-loss diets and the popularity of low-carb regimes. They spoke of feeling pressed for time to cook the things they wished they could cook.

We aspire to better food choices, yet the way we eat now is the product of vast impersonal forces that none of us asked for. The Amer-ican food system provides consumers with more than four thousand separate varieties of snack bar but only one banana, the Cavendish. The choices we make about food are largely predetermined by what's available and by the limitations of our busy lives.

It might be possible to eat in a more balanced way, if only we didn't have to work; or go to school; or save money; or travel by car, bus, or train; or shop at a supermarket; or live in a city; or share a meal with children; or look at a screen; or get up early; or stay up late; or walk past a vending machine; or feel depressed; or be on medication; or have a food intolerance; or own an imperfectly stocked fridge. Who knows what wonders we might then eat for breakfast?

It's now becoming abundantly clear that the way most of us currently eat is not sustainable—either for the planet or for human health. The signs that modern food is unsustainable are all around us, whether you want to measure the problem in soil erosion, in the fact that so many farmers cannot make a living from producing food, or in the rising numbers of children having all their teeth extracted because of their sugary diet. Food is the single greatest user of water and one of the greatest drivers of the loss of biodiversity. We cannot carry on eating as we are without causing irreparable harm to ourselves and to the environment. At some point, climate change may force governments to reform food systems to become less wasteful and more in tune with the needs of human health. The hope is— as we'll see—that some governments and cities are already taking action to create environments in which it is easier to feed ourselves in a way that is both healthy and joyous. In the meantime, many individual consumers have taken matters into their own hands and tried to devise their own strategies for escaping the worst excesses of modern food.

OUR CULTURE'S OBSESSIVE FOCUS ON A PERFECT physique has blinded us to the bigger question, which is what anyone of any size should eat to avoid being sickened by our unbalanced food supply. No one can eat themselves to perfect health, nor can we ward off death indefinitely, and the attempt to do so can drive a person crazy. Life is deeply unfair, and some people may eat every dark green leafy vegetable going and still get cancer. But even if food cannot cure or forestall every ill, it does not have to be the thing that kills us.

The greatest thing that we have lost from our eating today is a sense of balance, whether it's the balance of meals across the day or the balance of nutrients on our plate. Some complain that modern nutrition is in a state of terminal confusion and that science knows nothing about what a person should aim to eat for better health. This is not quite true. A series of systematic reviews of the evidence by some of the world's top nutrition scientists—the kind who are not funded by the sugary drink or bacon industries—have sifted through all the data and found robust causal evidence that regular portions of certain foods do significantly lower a person's risk of chronic diseases such as heart disease, diabetes, and stroke.[15]

It's the balance and variety of what you eat that matters rather than any one ingredient, but there are certain foods you might want to throw into the mix, depending on your preferences, your beliefs, your digestion, and whether you have a food intolerance or allergy. These protective foods are all relatively unprocessed and include nuts and seeds, beans and other legumes, and fish, the oilier the better (canned sardines are an affordable alternative). Fermented foods such as yogurt, kefir, and kimchi seem to help us in all kinds of ways that we are only starting to understand, from gut health to reductions in the risk of type 2 diabetes. There are also numerous benefits to eating foods high in fiber, especially vegetables and fruits and real whole grains (as opposed to almost any packaged food that is labeled as containing "whole grains"). You do not have to fork out for superfoods such as fashionable kale; any vegetables, and as many different types as possible, will do.

A good diet is founded less on absolutes than on the principle of ratio. Take protein. One of the missing links in the obesity crisis seems to be the falling ratio of protein to carbohydrate in our diets. This phenomenon—first documented in 2005 by biologists David Raubenheimer and Stephen Simpson—is known as the protein leverage hypothesis. In absolute terms, most people in rich countries get more than enough protein, much of it from meat. What has fallen, however, is the *proportion* of protein in our diets relative to carbohy-drates and fats. Because our food system supplies us with a flood of cheap fats and refined carbohydrates (including sugars), the percent-age of proteins available to the average person in the United States has dropped from 14–15 percent of total energy intake (which is fine for most people, assuming you are not a bodybuilder, but still on the low side) to 12.5 percent. This leaves many of us hungry for protein even if we have more than enough calories. Raubenheimer and Simpson have observed this protein hunger at work in many animal species besides humans. When a cricket is short of protein, it will resort to cannibalism. Locusts will forage different food sources until they get the ideal protein balance. Humans are neither as wise as locusts nor as ruthless as crickets. When our food is low in protein, we try to extract the balance from carbohydrates, resulting in overeating. If Rauben-heimer and Simpson are right, then obesity is—among many other things—a symptom of protein hunger.[16]

Protein leverage would also explain why low-carb diets work so well—at least in the short term—as a weight-loss tool for many peo-ple in our current food environment. The low-carb diet works in part because it is higher in protein (and lower in sugar). But there are other, gentler adjustments you could make to get your ratios back on track short of swearing off bread for life. You could cut down on sugary drinks, add yogurt or eggs to your breakfast, or go easy on carbs for just one meal a day. Or you could get more protein from green veg-etables and pulses, which turn out to be much richer in amino acids than was once believed.[17]

It isn't that there is anything wrong with carbohydrates per se (unless you are suffering from diabetes or insulin resistance). After all, humans have thrived on carbohydrate-rich diets in the past—and, as nutrition scholar David Katz remarks, carbs can mean anything "from lentils to lollipops." Our nutrient-obsessed age wants to fit every food into a certain box, yet legumes such as lentils are 25 percent carbohydrate and 25 percent protein. Do we welcome the lentil as a protein or reject it as a carb? Perhaps, instead, we should simply find a lentil recipe that tempts us to eat it (spiked with cumin seed and enriched with butter works for me) and call it food, because it is.

We are now at a transition point with food where a critical mass of consumers seem to be ready to make another set of changes to replace the last and, out of this craziness, to create new ways of eating that actually make sense for modern life. Very little about how we eat now would have been considered normal a generation ago, but I take consolation in thinking that surely much of it won't seem normal in the future either. From around the world, I have found hopeful signs that the pattern of our eating may be turning back again in a healthier and more joyful direction. In the final section, I celebrate some glimmers of a different food culture that is just emerging: one in which nutrition and flavor are finally joined up.

To reverse the damage being done by modern diets would require many other things to change about the world today, from the way we organize agriculture to the way we talk about vegetables. We would need to adjust our criteria of prosperity to make it less about money in the bank and more about access to good-quality food. We would need different food markets and differently run cities. Through education or experience, we would also need to become people with different appetites so that we no longer crave so much of the junk foods that sicken us. None of this looks easy at present, but neither is such change impossible. If the food changes we are living through now teach us anything, it is that humans are capable of altering almost everything about our eating in a single generation.

CHAPTER ONE

THE FOOD TRANSITION

THERE ARE TWO BIG STORIES TO TELL ABOUT FOOD today, and they could hardly be more different. One of these stories is something like a fairy tale; the other is closer to a horror story. Both, though, are equally true.

AND THEY NEVER WENT HUNGRY AGAIN

The happy version of the story goes something like this. Humans have never in history been fed as well as we are right now. As recently as the 1960s, you could go into almost any hospital in the developing world and find children suffering from something called kwashiorkor, a form of severe protein malnutrition that gives rise to swelling all over the body and causes a potbelly. Now, kwashiorkor is mercifully rare in most countries (though it still afflicts millions in central Africa). Other diseases of deficiency, such as scurvy, pellagra, and beriberi, are—with a few exceptions—terrors of the past. The waning of hunger is one of the great miracles of modernity. *And they never went hungry again* is the happy ending of many fairy tales.[1]

Until the twentieth century, the threat of famine was a universal aspect of human existence across the world. Harvests failed;

populations starved. For anyone but the wealthy, food wasn't something to be relied on. Even in rich countries, such as Britain and France, ordinary people lived with the daily specter of going to sleep hungry and spent as much as half their income on basic staples such as bread or grains. In the rice-based economies of Asia, mass starvation regularly killed whole communities.

In 1947, half of all the people on the planet were chronically underfed, according to the Food and Agriculture Organization (FAO) at the United Nations. By 2015, that figure had dropped to one in nine—even though the overall population had risen astronomically during the same period. The number of people living in extreme poverty continues to decline dramatically. On any given day in 2017, the numbers affected by extreme poverty—defined as less than $1.90 a day per person to cover food, clothing, and shelter, adjusted for inflation—declined by 250,000.[2]

Absolute hunger is much rarer than it once was. In 2016, the Swedish historian Johan Norberg went so far as to argue, in his book *Progress*, that the problem of food had been solved. Advances in farming technology over the course of the twentieth century made massively more food available to vastly more people. A modern combine harvester can harvest in six minutes what it once took twenty-five men a day to do, and modern cold storage can prevent crops from rotting and being wasted after harvest.[3] More food is produced each year than ever before.

Perhaps the greatest changes of all came about through the invention in the 1910s of the Haber-Bosch process, a method for synthesizing ammonia that made highly effective nitrogen fertilizers cheap to produce for the first time. Vaclav Smil, a Canadian expert on land use and food production, has calculated that as of 2002, 40 percent of the world's population owed their existence to the Haber-Bosch process. Yet how often do you hear anyone talking about Haber-Bosch? Without it, many of us might not be here today, yet it has far less name recognition than Häagen-Dazs, a supposedly

Danish label for a brand of ice cream dreamed up by a businessman in the Bronx in 1961. In a way, our ignorance about Haber-Bosch shows once more how lucky we are. We have reached the point where most of us can afford to think more about ice cream than about survival.[4]

It is said that Norman Borlaug, a plant agronomist who was awarded the Nobel Peace Prize in 1970, saved a billion lives from starvation with his invention of semi-dwarf high-yield wheat varieties. Thanks to Borlaug's miracle wheat—coupled with modern farming methods—yields of the crop nearly doubled in India and Pakistan from 1965 to 1970.

Many of us yearn for the good old days of food, when it was normal to bake your own bread—or roll your own tortelloni, as the case may be—but no one would wish themselves back to a state of famine. We sometimes forget that for most of history, even in rich countries life expectancy was short and people were sometimes so deprived of food that they mixed tree bark into flour to make it go further. Even for those who did not suffer actual famine, the business of cooking and eating on an average family budget could lead to a pinched and frugal existence, especially in winter, when—before refrigeration was available—meals centered on staple grains and salted meat with little that was green or crunchy, never mind spicy or particularly delicious.[5]

Today, many of us have instant access to almost preposterous quantities of food, year-round, of a freshness and variety our grandparents could not have imagined. In the city where I live, a three-minute walk from my home in any direction will take me to food shops with plentifully stocked shelves. I can stroll east and arrive at a Chinese supermarket, a butcher, and a South Asian grocery that sells everything from fresh mint leaves and every spice under the sun to homemade falafels and samosas. To the north, I will find a health food co-op offering local sourdoughs, ancient grains, and organic apples and a Hungarian deli selling any European cheese I can possibly name, as well as a few that I can't. To the west and the south are four rival supermarkets, each heaving with fresh fruits and cereals, meat and fish, oils and vinegars, ginger and garlic.

Magical as it is, I've come to feel entitled to this abundance. On the rare occasions that I arrive at one of these many shops and the one specific thing I was expecting to buy has run out—no parmesan left on a Sunday night! Outrageous!—I feel a mild consternation, because my expectation that I can eat exactly what I want at the precise moment I want to eat it has been scuppered.

In the developed world, many are living in a new age of delicious, liberated from the last vestiges of postwar austerity. The decline of hunger has been accompanied by a bright new dawn of flavor. Cooks are relearning the arts of pickling and fermenting, but this time, we are doing it out of love, not necessity. Never have so many cups of heavenly tasting coffee been topped with so many variations on beautiful latte art. Clever home cooks have made food far more inventive and open. Gone is the old food snobbery that said you couldn't be a good cook if you hadn't mastered half a dozen elaborate French sauces or a shellfish bisque. The internet has enabled recipe swapping on a scale and at a speed that is dizzying. Where our grandparents (in the Anglo-American world at any rate) sat down dutifully to plates of underseasoned meat and two vegetables, we have developed unexpected new global palates: for spicy Turkish eggs sprinkled with sumac or vibrant salads of green mango and lime. Food has gone from being a scarce and often dull kind of fuel to an ever-present, flavorsome, and often exotic experience, at least in big cities. Think how casually we eat ingredients such as kalamata olives or couscous now, as if born to them.

Yet the omnipresence of food has created its own completely new difficulties. Widely available cheap food can look like a dream—or it can be a nightmare. It's impossible to accept Norberg's assertion that the problem of food has been solved when diet now causes so much death and disease in the world. The same food that has rescued us from hunger is also killing us.

As of 2006, for the first time the number of overweight and obese people in the world overtook the number who were underfed, in absolute terms. That year, eight hundred million individuals still did not have enough to eat, but more than one billion were overweight or obese. To our hungry ancestors, having too much to eat might have looked like the gold at the end of the rainbow, but what these new calories are doing to our bodies is not a happy ending.[6]

The problem isn't just that some people are overfed and others are underfed, lacking enough basic calories to ward off gnawing hunger (though that still remains a real and brutal problem). The new difficulty is that billions of people across the globe are *simultaneously overfed and undernourished: rich in calories but poor in nutrients.* Our new global diet is replete with sugar and refined carbohydrates yet lacking in crucial micronutrients such as iron and trace vitamins. Malnutrition is no longer just about hunger and stunting; it is also about obesity. The literal meaning of *malnutrition* is not hunger but bad feeding, which covers inadequate diets of many kinds. If governments have been slow in acting to tackle the ill health caused by modern diets, it may be because malnutrition does not look the way we expect it to.

Despite the decline in hunger, malnutrition in all its forms now affects one in three people on the planet. Plenty of countries—including China, Mexico, India, Egypt, and South Africa—are suffering simultaneously from overfeeding and undernutrition, with many people suffering from a surfeit of calories but a dearth of the crucial micronutrients and protein a body needs to stay healthy. As a result, not just in the West but across the world, people are suffering in growing numbers from diseases such as hypertension and stroke, type 2 diabetes, and preventable forms of cancer. The lead cause of these diseases is what nutritionists call "suboptimal diet" and what to the rest of us is simply "food."[7]

Our ancestors could not rely on there being enough food. Our own food fails us in different ways. We have markets heaving with bounty, but too often what is sold as "food" fails in its basic task: to nourish us.

To walk into the average supermarket today is to be greeted by not just fresh, whole ingredients but also aisle upon aisle of salty, oily snacks and frosted cereals, of bread that has been neither proofed nor fermented, of sweetened drinks of many hues, and of supposedly healthy yogurts that are more sugar than yogurt. These huge changes to modern diets have gone hand in hand with other vast social transformations, such as the spread of cars, electric food mixers, and electronic screens of many kinds, which have left us far less active than earlier generations, gym memberships or not. The mechanization of farmwork that created the food to feed billions also resulted in farmers (in common with most of the rest of us) leading increasingly sedentary lives.

In just a few decades, these alterations to how we eat have left unmistakable marks on human health. Take type 2 diabetes. The causes of this chronic condition, whose symptoms include fatigue, headaches, and increased hunger and thirst, are still being debated by scientists, but there is clear evidence that—genetics aside—there is a higher risk of getting type 2 diabetes if you habitually consume a diet high in sugary drinks, refined carbohydrates, and processed meats and low in whole grains, vegetables, and nuts.[8]

In 2016, more than six hundred children in the UK were registered as living with type 2 diabetes. Yet as recently as 2000, not a single child in the country suffered from the condition.[9]

So are we living in a food paradise or a food hell? It doesn't seem possible to reconcile these competing stories about modern food. But in 2015, a group of scientists in the United States, the UK, and Europe devised a systematic assessment of the world's diet that showed that both stories are true: the world's diet is getting better and worse at the same time.

WHERE THE BALANCE FALLS

The light is fading on a cold winter's day. I am sitting in a café at the top of the Cambridge University graduate students' union with

Fumiaki Imamura, a thirty-eight-year-old scientist. He drinks black coffee; I drink English Breakfast tea. Imamura, who has a Beatles haircut and a bright purple tie, is originally from Tokyo but has spent the past fifteen years in the West, studying the links between diet and health. "There are so many myths about food," Imamura says. One of the myths he refers to is the notion that there is such a thing as a perfectly healthy diet.

Every single human community across the globe eats a mixture of the healthy and the unhealthy, but the salient question is where the balance falls. Imamura's research shows that most countries in the world are currently eating more healthy food than we ever did but also more unhealthy food. Many of us have a split personality when it comes to food, but then this is hardly surprising given how schizophrenic our food supply has become. We have access to more fresh fruit nowadays than we ever did but also more sugar-sweetened cereals and french fries.

Imamura is a nutritional epidemiologist, meaning that he studies outlines of diet across whole populations to arrive at a more accurate account of how food and health are related. He works in the MRC Epidemiology Unit on the Cambridge Biomedical Campus. Imamura is one player in a much larger research team that straddles multiple universities in the United States and Europe. The overall project is based at Tufts University in Boston and is led by Professor Dariush Mozaffarian, one of the leading scholars currently using big data to measure nutrition in countries worldwide.

In 2015, Imamura was the lead author on a paper in the medical journal *The Lancet* that caused a stir in the world of nutrition science. This team of epidemiologists has been seeking to map the healthiness, or otherwise, of how people eat across the entire world and how that health has changed in the twenty years between 1990 and 2010.[10]

At this point, you might ask, what counts as a good-quality diet? Some would define healthy food in positive terms: how many vegetables and portions of oily fish a person eats. Others define it more negatively, judging it by an absence of sugary drinks and junk food. Clearly, these are two very different ways of looking at the question. Most research on diet and health has lumped the two together, assuming that a high intake of "healthy" fish will automatically go along with a low intake of "unhealthy" salt, for example. But, alas, human beings are inconsistent creatures.

The Japanese, who are generally considered to eat an outstandingly "healthy" diet as rich nations go, consume large amounts of both fish and salt: the one "healthy" and the other "unhealthy." They consume much refined polished white rice ("unhealthy") along with copious amounts of dark green vegetables ("healthy"). Imamura himself still eats a diet centered on vegetables and fish, he tells me, but also a lot of salt in the form of soy sauce, even though as an epidemiologist he is aware that high sodium intake has been linked in numerous studies to high blood pressure. But he is conscious that no population in the world eats exactly the combination of healthy foods that nutritionists might recommend.

There have been many attempts to measure the healthiness of the world's diet in the past, but most studies have treated human eaters as more rational than we actually are. Previous studies have summed together high consumption of healthy foods and low consumption of unhealthy foods. What made Imamura's paper so innovative—and so much closer to the way we actually behave around food—was that he and his fellow researchers studied healthy and unhealthy foods in two parallel datasets.

Imamura and his colleagues came up with a list of ten healthy items: fruits, vegetables, fish, beans and legumes, nuts and seeds, whole grains, milk, total polyunsaturated fatty acids (the kind of fat found in seed oils such as sunflower), plant omega-3s, and dietary fiber. They created a separate list of unhealthy items: sugary beverages, unprocessed red meats, processed meats, saturated fat, trans fat, cholesterol, and sodium. (Imamura knows that some would quibble with the items on these lists. There is an ongoing debate among nutrition scientists about the healthiness or otherwise of saturated fats. It looks as though the key question with saturated fat, as with other nutrients, is not whether it is unhealthy in absolute terms but what you choose to eat instead of it. There is evidence that replacing saturated fat with processed carbohydrates can be harmful for heart health, whereas replacing it with olive oil or walnuts has benefits.[11] But based on everything that the epidemiologists currently knew about patterns of diet and health outcomes, these lists were the best they could do.) The researchers then tried to map a pattern of how much of these healthy and unhealthy foods are eaten in any given country.

"We don't know very much about what people consume, actually," Imamura tells me, disarmingly, sipping his black coffee. "Assessment of diet is very difficult." Almost all the data we have on what people eat is based on market figures: what commodities come into the country, or how many packets of something people buy in any given year. This data on supply and production is used as a proxy for what people actually eat. It is useful for mapping big changes in our diets over time—the rise of salmon and the fall of herring, say. Often, food supply data reveals big truths about what we eat that are invisible to us in the daily bustle of shopping and cooking. Much of what I'll tell you about food in this book will come from market data because often it's the only hard data available.

But this kind of market data has flaws. For one thing, it offers only a national average, and for another, it does not tell you what happens to the food after it enters the home. Did the consumer steam that bag of green beans and eat them with grilled sardines? Or leave it to rot at the back of the fridge?

Another method of measuring diets is to ask people what they eat, whether over a twenty-four-hour period or in a seven-day diary. Imamura tells me he much prefers survey data to market data because it gives a more detailed picture of how consumers actually behave around food. The snag is that one of the ways we behave around food is that we lie about it: No, I never bought and ate those extra-cheesy nachos. Yes, I eat five fruits and vegetables a day, every day. We also forget things, like that candy bar we devoured in haste between meetings.

One way to get around this problem of accuracy is to measure biomarkers in the human body itself, like forensic scientists analyzing a corpse. In recent years, epidemiologists have started searching for traces of our diets in blood serum, hair samples, and even toenail clippings (toes are used instead of fingers because they are less exposed to outside environmental contamination). Toenail clippings are apparently the best way to measure levels of the mineral selenium in the body, which is something nutrition researchers are interested in, since low selenium correlates with type 2 diabetes and childhood obesity.

The most versatile and commonly used biomarker to determine dietary intake is urine. Unlike toenails, which take weeks to grow back, urine is—how to put this delicately?—endlessly renewable, and it reveals traces of more different foods than any other measure. We haven't quite reached the point yet where a sample of your urine could tell a researcher that you ate spinach gnocchi for lunch and pumpkin risotto for dinner, but that day may not be far off. In the meantime, urine has most often been used to measure how much salt we eat. Imamura and his colleagues looked at 142 surveys that measured sodium levels in urine, providing data on salt consumption for the majority of adult humans on the planet.[12]

At the time of writing, Imamura's study is the most complete snapshot we have of diet quality on a truly global scale as it relates to patterns of ill health. In all, the researchers managed to find data to cover 88.7 percent of the adult population of the whole world. From this, they built up a picture of what we eat from two different angles: on the one hand, how much healthy food countries eat and, on the other hand, how much unhealthy food.

A person may enjoy eating a slice of fresh melon but also enjoy munching on greasy fried onion rings. Countries, too, have contradictory tastes. Since 1990, the planet's consumption of healthy items has undoubtedly been going up, but this does not mean that people necessarily have a healthy pattern of eating. Take fruit. Since 1990, world vegetable consumption has remained static, but the world's fruit intake seems to have gone up by an average of 0.2 ounces per person per day. For people who can afford to buy it, fresh fruit, from grapes to watermelon, has become one of the world's favorite snacks. Fruit is expensive, and it's one of the first things parents buy as a treat for their children when they start to have disposable income. The rise of fruit gives credence to the fairy story about modern food (setting aside the fact that modern fruit is often not as nutritious as fruit used to be). Out of 187 countries, all but 20 or so have increased their intake of healthy foods, especially foods such as fruit and unsalted nuts that are eaten between meals.[13]

But Imamura's paper also supports the food horror story. The data clearly shows that diets high in sugary drinks, trans fats, salt, and processed meats became much more common in the world between 1990 and 2010. In 2010, around half the countries in the world were eating a diet higher, often drastically higher, in unhealthy items than in 1990. The prevalence of unhealthy items in our diets is increasing more rapidly than our consumption of healthy foods. But it is not increasing everywhere to the same extent.

The biggest surprise to come out of the data was that the highest-quality overall diets in the world are mostly to be found not in rich countries but in the continent of Africa, mostly in the less developed sub-Saharan regions. The ten countries with the healthiest diet patterns, listed in order with the healthiest first, came out as follows:

Chad	Sierra Leone
Mali	Laos
Cameroon	Nigeria
Guyana	Guatemala
Tunisia	French Guiana

Meanwhile, the ten countries with the least healthy diet patterns, listed in order from the bottom up were:

Armenia   Iceland
Hungary   Latvia
Belgium   Brazil
United States   Colombia
Russia   Australia

The idea that healthy diets can be attained only by rich countries is one of the food myths, Imamura says. He found that the populations of Sierra Leone, Mali, and Chad have diets that are closer to what is specified in health guidelines than those of Germany or Russia. Diets in sub-Saharan Africa are unusually low in unhealthy items and high in healthy ones. If you want to find the people who eat the most whole grains, you will either have to look to the affluent Nordic countries, where they still eat a lot of rye bread, or to the poor countries of southern sub-Saharan Africa, where a range of nourishing grains such as sorghum, maize, millet, and teff are made into healthy main dishes usually accompanied by some kind of stew, soup, or relish. Sub-Saharan Africa also does very well on consumption of beans, pulses, and vegetables. The average Zimbabwean eats 17.3 ounces of vegetables a day, compared with just 2.3 ounces for the average person in Switzerland.[14]

It was Imamura's conclusion about the high quality of African diets that ruffled feathers in the world of public health. What about African hunger and scarcity? Zimbabweans may eat more vegetables than the Swiss, but there is more to health than vegetables, given that life expectancy in Zimbabwe in 2015 was just fifty-nine years of age compared with eighty-three for the average Swiss person. Some scientists argue that the low score for unhealthy foods in some African and Asian countries is actually a sign of diets that are "poor" in various ways. If the people of Cameroon consume low amounts of sugar and processed meat, it is partly because they are consuming low amounts of food all round.[15]

Imamura does not deny, he tells me, that the quantity of food available is very low in some of the African countries but adds, "That's not the point of our study. We were looking at quality." His paper was predicated on the assumption that everyone in the world was consuming two thousand calories a day. Imamura was well aware that is far from the case in sub-Saharan Africa, where the prevalence of malnourishment is around 24 percent according to the Food and Agriculture Organization. But he and his colleagues wanted to isolate the question of food quality from that of quantity. Traditional public health nutrition, he observes, was so fixated on the question of hunger that it paid too much attention to the quantity of food people had access to without considering whether the food itself was beneficial for human health.[16]

Africa's hunger can easily blind us to the sheer quality and variety of food that people enjoy in much of the continent. The findings of Imamura's paper came as no surprise to Graeme Arendse, a South African journalist at the *Chimurenga Chronic*, a magazine celebrating pan-African culture. In 2017, Arendse helped put together a special food issue of the magazine that challenged the Western idea that African food was all about deprivation and suffering. On a sunny winter's day, sitting in his offices in Cape Town above the pan-African market in the city center, Arendse tells me that "this story of scarcity is not true." Arendse sees traditional African food as deeply diverse, with much of it very healthy. A short walk from his office in Cape Town, Arendse can pick up takeout of fish and brown rice at a Malian place where he likes to go. Other days, when the mood hits, he goes to another café to buy a bowl of Nigerian egusi soup made from melon seeds with seafood and bitter greens for the same price as a fast-food meal from McDonald's.

Arendse worries that unless traditional African cuisine with its soups and stews of many kinds is celebrated more, it will lose out even more to the fast foods and convenience foods that he notices becoming so popular now in South Africa. On the bus into work, in just the past couple of years, he has started to see some commuters breakfasting on crisps and cans of cola. "I never saw that in the past."

Dietary patterns are getting rapidly worse in much of Africa, including South Africa. In recent years, moneyed South Africans have abandoned the old dinners of mealy maize and have started to drink bottles of sparkling mineral water and to eat salads of roasted vegetables and feta cheese, as well as, yes, many kinds of avocado toast. But there has also been a colossal rise in the consumption of packaged snack foods and sugary drinks. The balance of what South Africans eat is tipping away from the old vegetables and stews of the rest of sub-Saharan Africa and toward a Westernized diet of fried chicken, burgers, and oversized portions of pasta.[17]

"These young people have stretched their stomachs," observed an old black South African, speaking to a dietitian in 2016, startled by the way that children suddenly expected to eat fried foods and meat every day.[18] Middle-income countries such as South Africa have experienced the full fairy tale and the full horror story of food at the same time. Rates of both undernutrition and overnutrition in South Africa exceeded 30 percent of the population as of 2016. In the old days, South Africans ate many wild fruits and breakfasted on a thick maize or sorghum porridge, seasoned to taste with a few drops of vinegar. Now, breakfast is more likely to be nutrient-poor white industrial bread with margarine or jam. With escalating sugar consumption, tooth decay is rising in South Africa at an alarming rate.[19]

Eating in South Africa, a parched land with relatively poor soil quality, has never been "heaven on earth," as South African dietitian Mpho Tshukudu has written. There is no golden age of food to return to. But neither have South Africans ever had to face food dilemmas quite like the ones they face today on a daily basis. One mother in her forties who came to Tshukudu's clinic recalled that as a child growing up in a rural village, she walked for miles and ate home-cooked foods every day, always with a vegetable or some kind of legume. She knew no one who was obese and never needed to visit a doctor. But now, this woman lived in the city with her husband and three children, and they all ate a lot of takeout food and were frequently unwell. Her nine-year-old daughter was already so big that, to her distress, she had to buy her clothes in the grown-up section of the store.[20]

In some ways, South Africa's new unhealthy pattern of eating is distinctive to the country itself and to the injustice of the apartheid years. During apartheid, the state controlled who moved to towns and who stayed in the country, and no black farmers were allowed to own land outside the homelands. Adults living in black townships often had long commutes to jobs in the white cities, which left less time for cooking than in the past, and as a result, some of the old traditional dishes died out.

But the most extreme and sudden changes to South African eating happened after the end of apartheid in the mid-1990s, during and after Nelson Mandela's presidency, when thousands of black South Africans were lifted out of poverty for the first time. People were free to move to the cities, and they did. By many metrics, life got better and easier, but much of what people began eating was less healthy than it had been before. As a newly open economy, the country was flooded with fast food and processed food from both home and abroad. From 2005 to 2010, the sales of processed snack bars in South Africa increased by more than 40 percent.[21]

New freedom and city living, new snacks and abundance, new obesity and type 2 diabetes: the patterns of both eating and health have shifted fast in South Africa since the 1990s. The speed at which diets are changing here is vertiginous, yet the pattern is a familiar one. It is almost as if South Africa—along with so many other countries in the world—is following a script for eating set by America fifty or so years ago.

STAGE FOUR

Growing up in 1950s Wisconsin, Barry Popkin drank only tap water and milk, except for a small glass of orange juice to start the day. His father drank tea, and his mother had coffee. At the weekend, as he has explained in his 2009 book *The World Is Fat*, his parents might take a glass of wine for a treat. No one in Popkin's family drank sweetened lattes or sugary energy drinks, and the adults would not have dreamed of drinking alcohol every day. There were no smoothies and no white chocolate mocha Frappuccinos. Popkin—professor of nutrition at the University of North Carolina–Chapel Hill—has made it his life's mission to study the reasons why our patterns of eating and drinking are so different from those of the past and to figure out ways to save the best of the changes and move beyond the worst of them.[22]

During the months when I was first researching this book, it felt as if all roads led to Barry Popkin. Whether I was looking for hard facts on snacking or sugar or statistics about how food had changed in China over the past decade, Popkin always seemed to have coauthored the definitive paper on the subject. He was also involved in working with governments to create better food policies in many countries, including Mexico, Chile, Colombia, and Brazil. His website showed a photo of a cheery-looking man in his seventies with a white beard, but this Popkin was so prolific, I was starting to doubt whether he really existed or whether he was in fact a team of nutrition academics working out of a factory somewhere.

When I contacted Popkin to arrange a telephone interview, he emailed straight back and told me he was having a "horrendous" week but could take my call at 9:00 a.m. EST precisely on Monday morning. A gruff-voiced man answered the phone and immediately started explaining how food had radically changed in recent years, not just for a few people, but for billions across the world. He spoke with great authority about the marketing of chips and convenience foods and about the rise of highly sweetened drinks and the fall of home cooking. "It's a radical change," Popkin told me, "and it's going to be a big battle to reverse it."

Popkin's interest in nutrition started, he has written, during a year in India in 1965–1966 when he was an economics student living in shantytowns in Old Delhi. India was a shock because after his modest but comfortable American childhood, Popkin was exposed to the extremes of hunger firsthand. He returned to the States determined to use economics to help improve the way people ate. He assumed at the time that the great problem for nutrition to solve would always be hunger.[23]

By the 1980s, however, Popkin had noticed that obesity had begun to replace hunger as the main nutritional problem in the Western world, and he observed, aghast, as the same set of chronic diseases swept across the globe. He was one of the first experts in the field to argue that obesity was a global problem, not a phenomenon of the West. Popkin coined the phrase "nutrition transition" to explain the changes he saw happening around the world as countries developed from poverty to riches. As a country becomes richer and more open to global markets, its population almost inexorably starts to eat differently, consuming more oil and meat and sugar and snack foods and fewer whole grains and pulses. Wherever this diet was adopted, Popkin noticed, it brought with it easier lives as well as a host of diseases.[24]

One way to think about human history is as a series of diet transitions, with each stage driven by changes in the economy and society, as well as shifts in technology, climate, and population. In the beginning, we were hunter-gatherers, eating a mostly low-fat diet of varied wild greens, berries, and wild animals. During the Upper Paleolithic period, which began about fifty thousand years ago, more than half of our food came from plants, and the rest came from animals. In these societies, people were forced to collaborate to collect food. We had discovered fire but not cooking pots. Life expectancy was low—you were at risk of dying a violent death if infectious diseases didn't get you first. But the archaeological record suggests that (depending on where in the world they lived) the humans in this phase who survived into adulthood experienced mostly good health, with few nutritional deficiencies.

Stage two, starting around 20,000 BCE, was the agricultural age, which was characterized by a switch to staple cereals and a huge increase in population. Now we had clay cooking vessels and more sophisticated grindstones at our disposal. The hunter-gatherer diet of wild plants and meats gave way to diets based on staple cereals, whether it was the rice and millet of China or the barley of Mesopotamia. Farming bestowed huge benefits. It created food surpluses for the first time, which freed many people from the task of food gathering and gave rise to vast new civilizations such as that of the Indus Valley, where modern-day Pakistan lies. Grains were a very efficient way to generate calories from the land. Without agriculture, there would have been no cities, no politics, and no human civilization as we know it.

The downside of farming, however, was that it gave people a less varied range of foods than before. Along with the adoption of staple cereals, phase two saw a rise of famine and a sudden increase in diet-related problems. With diets that were often inadequate both in quantity and quality, humans shrank in stature and suffered from a range of deficiency diseases. The difference in human health between the diets of stage one and stage two is the rationale behind the popular Paleo diet, in which modern dieters try to turn the clock back by ten thousand years or so and eat as if farming had never been invented.

Then again, to find a diet healthier than the one most people eat today, we don't need to go back thousands of years. In Europe, we could go back a mere couple of hundred years to the third stage, which Popkin calls "receding famine." During this period, advances in agriculture, such as crop rotation and fertilizer, led to a more varied and plentiful diet, with fewer starch-based staples and a bigger variety of vegetables along with animal protein. In stage three, the possibilities of cooking expanded, with new methods of drying, preserving, and pickling. This period also witnessed a slow decline in mortality. Many of the old deficiency diseases—such as scurvy and beriberi—became less common as diets became more nourishing. On Popkin's model, many sub-Saharan African countries are living through this stage now. This would explain why their diets compare so favorably, in Imamura's paper, with those of the industrialized world.

But then comes stage four, which is where we are now. This era is different in quality from any of the other stages. Suddenly, the diet changes much more rapidly, with consequences for human health that are more extreme. The economy shifts away from manual labor and toward mechanization, people move from the countryside to cities, and they start to expend less energy. There are revolutions in food processing and marketing, and people start to eat more fat, more meat, and more sugar, with far less fiber. Stage four sees human life expectancy hit new highs with the decline of deficiency diseases and the wonders of modern medicine. But it also sees populations suffering from diet-related chronic illness as never before. This "nutrition transition" happened all over the Western world in the decades after the Second World War and is now happening even faster among low- and middle-income nations in the rest of the world. This transition explains why our food is sickening us now, through excess rather than hunger.

Stage four is a radical break with the past that represents a reinvention of food and what it means for human life. One of the greatest departures of stage four is the new homogeneity of food. As agriculture becomes a vast international form of trade, people start relying on the same small number of global crops as each other, even when they live oceans and continents apart.

For centuries, eaters have marked high days and holidays with moreish fried foods such as fritters and doughnuts. Only in modern times, however, could a person buy a stackable carton of fried chips made from a slurry of dried potatoes and wheat starch seasoned with barbecue flavoring and sit on a sofa eating them, not for a celebration, not even out of hunger, but just out of a mild feeling of restless boredom. Only in stage four could another person—in the same mildly bored state—be eating exactly the same chips at the exact same moment on another sofa somewhere halfway across the world.

THE GLOBAL STANDARD DIET

The nutrition transition has not just taken place at the level of supply. It has also altered our personal hungers so that we become people who—to a bizarre extent—gravitate toward the same foods. Between the 1960s and today, people around the world stopped depending so much on their own particular foodstuffs, the ones that belonged to our own families and homelands, and started eating other, alien commodities, grown in faraway places. Soon, we were eating so many of these alien foods that they stopped tasting strange to us and started tasting normal. We changed not only the dishes we ate but also the basic composition of our diets.

Nations have adjusted their food habits many times before—after all, tomatoes are not native to Italy, nor is tea to Britain—but the recent global homogenization of taste is unprecedented. All at once, billions of eaters in disparate places have started eating from the same repertoire of ingredients. Never before has such dietary change happened on such a scale and simultaneously across most of the planet. It is a switch so pervasive and so huge that we haven't had time to react or even to notice exactly what has changed. It is as if the color of the sky morphed from blue to green, but before we could protest that something was not right, our eyes adjusted, and we carried on as normal.

In the past, it was a fundamental fact about human beings—and about food—that people ate different things in different places. It's in our nature as omnivores to be skilled at adapting to varied food environments. If you ask someone, "What's food?" you would expect to receive wildly different answers to the question whether you were in Lagos or in Paris. In the past "food" was not one thing but many, varying according to local crops, local ingredients, and local ideas and prejudices.

When I was a child in the 1980s, I remember grown-ups in Britain talking with horror about the fact that the Japanese liked to eat raw fish! It seemed so improbable. From their tone of baffled revulsion, these Britons might as well have been contemplating swallowing live frogs. I never imagined that one day those same grown-ups, older and grayer, would stroll into a perfectly normal shop on the average British high street and casually pick up a tray of sushi for lunch. We now live in a clone-world where you can get pizza in Beijing, Chinese dumplings in Rome, and sandwiches everywhere and not even be startled by the incongruity.

At a cultural level, some of this change has been wonderful to see (and to eat). So many of the old barriers and prejudices that kept people from experiencing each other's food have been ripped apart. Many Westerners who used to look with suspicion on anything too garlicky or spicy or strong will now happily eat Korean-spiced barbecue or fiery Thai curries.

But if our palates have widened in some ways, they have narrowed in others, particularly at the level of ingredients themselves. When food becomes a common language across the whole planet, it stops being food at all, as our ancestors would have understood it. No matter where on the planet we live, there's a striking convergence going on in our eating habits.

In the early 2010s, a team of researchers led by Colin Khoury, an American plant diversity expert, set out to quantify how the world's diet had changed over the past roughly fifty years from 1961 to 2009, using food supply data from the FAO. For every country they could gather evidence for (152 of them, representing 98 percent of the world's population), they measured which crops were eaten and how many per capita calories and other nutrients each of the foods delivered. Overall, the researchers looked at fifty-three different foods, from oranges to rice and from sesame seeds to corn.[25]

These researchers found that there had been massive changes in eating since the 1960s. Wherever in the world you happen to live, you will now have access to much the same menu of core ingredients as someone who lives a thousand miles away in any direction. Khoury's team referred to this phenomenon as the Global Standard Diet.[26]

I started scrolling through the data on the FAO website trying to ascertain how the average global eater in the 1960s differed from the average eater today. Then I realized the very question I was asking was wrong. The whole point is that in the 1960s, there was *no such thing* as an average eater across most countries, just lots of specific and wildly divergent patterns of eating. Back then, there were maize eaters in Brazil and sorghum eaters in Sudan. There were steak-and-kidney pie enthusiasts in Britain and goulash devotees in Hungary. But it made little sense to ponder how a globally average person might eat because no such person existed.

It is only now that we can speak of a Global Standard Eater, because it is only now that humans have come to eat in such startlingly similar ways. Perhaps the biggest change is in the quantities that we eat—around 500 calories on average more than our equivalents in the 1960s (from 2,237 calories in 1961 to 2,756 calories in 2009). The Global Standard Eater consumes a whole lot more of almost everything than most eaters of the past. From the 1960s, we started to eat more protein and more fat, we drink more alcohol, and quite simply, we eat much more food. The average eater consumes a lot of sugar and rice and very few pulses or beans. Our diets overall are becoming sweeter and fattier and meatier, and we are highly dependent for our sustenance on foods that have been grown or produced far away from the place where we live, wherever that place might be. Khoury and his colleagues have calculated that more than two-thirds of national food supplies across the world are derived from crops that are foreign to the country where they are eaten.[27]

One gray rainy spring morning, I am talking with Colin Khoury over the phone. He is at his home in Colorado, where he works at the US National Seedbank. His background, he explains, is not in nutrition but in plant science. "I'm a diversity person," he says—one of the many biologists who believe that the future of the planet depends on maintaining the maximum biodiversity for healthy ecosystems. As he and his colleagues began to draw together all the data on the world's food supply, Khoury was startled to see just how homogenous the global diet had become, with eaters tending toward a common mean.

In Denver, where Khoury lives, the breakfast burrito is a local favorite in diners and cafés, especially on weekends. This greasy and comforting wrap is made from flour tortillas stuffed with eggs, potatoes, green chilies, maybe cheese, and some kind of meat—sometimes chorizo, sometimes bacon or steak. These sandwiches are an object of local pride, like the Philadelphia cheesesteak.

To those who love it, the Denver breakfast burrito is a distinctive thing. But in another sense, this local American specialty is not local at all. The bacon and the eggs in this Denver sandwich come from giant production lines in Iowa. The eggs are fried in soybean oil from Brazil. As for the wheat that makes the tortilla that binds the whole meal together, it is the same dusty refined white flour made from the same flavorless modern strain of wheat that goes into most breads in America, from bagels to sliced Wonder Bread to hotdog buns. The ingredients may be shuffled differently, but the Denver breakfast burrito is built from much the same deck of cards as a New York hamburger and fries or a pepperoni pizza in the Philippines.

"People are eating much more of the same crops," Khoury tells me. "We have all these local twists on food, but underneath it is not a huge list of species." In a way, the leap into stage four is like the emergence of agriculture in stage two: a narrowing of the diet that brings new diseases in its wake.

When you strip away the packaging, the recipes, and the brand names, most humans—from Rio to Lagos—are getting a sizeable majority of our energy from meat, sugar, refined wheat, rice, and refined vegetable oil. The average global eater largely consumes certain staple items, most of which will have been internationally traded before they reach the shop or the plate. The average eater gets the bulk of his or her daily calories (1,576) from just six sources. These are:

animal foods   sugar
wheat   maize
rice   soya beans

Of these, animal foods and wheat each contribute around five hundred calories, with a further three hundred calories apiece coming from rice and sugar, two hundred calories from maize and seventy-six calories from soya beans. Compared to these big six items, all the other food commodities pale into insignificance.[28]

There has been a startling shift away from multiple traditional diets toward a single modern one, with the same sweet-salty flavors and the same triumvirate at its heart of rice, wheat, and meat.

You can trace the effects of these homogenous diets all the way to the gut. Compared to the average affluent Westerner, a hunter-gatherer from the Hadza tribe in north-central Tanzania—subsisting on an ever-changing diversity of roots, berries, and wild meats—has 40 percent more microbiome diversity (the microbiome being the host of micro-organisms in the human gut). Having a less diverse gut microbiome has been linked with both obesity and type 2 diabetes.[29]

It's worth noting that in some countries the move toward a global average diet has been beneficial. "In some places," Khoury points out, "it actually means an increase in diversity," certainly compared to fifty years ago. Averaged out, the world's diet is more balanced now than it was in 1960, if balance is defined as eating an even spread of different foods. Until recently, many countries in East Asia were dangerously dependent on the single staple of rice to feed themselves. Apart from being a monotonous way to live, such single-staple diets are precarious when the single crop happens to fail—as the Irish potato famine demonstrated in the nineteenth century. Thanks to the opening up of new global markets, East Asian countries, such as Vietnam, have now been able to diversify into wheat and potatoes and as a result bestow greater food security as well as more varied nutrients.[30]

But in most places, the new global diet has involved a narrowing down of what people eat. Our world contains around seven thousand edible crops, yet 95 percent of what we eat comes from just thirty of them. As omnivores, humans are designed to eat a varied diet, so there's something strange and wrong when, as a species, we become so limited in our choice of foods.[31]

It might surprise you (as it did me) to learn that the most average place in the world, foodwise, is not the United States, which is actually pretty extreme in the composition of its diet. To take one example, Americans have access to around *twice* the global average calories from meat (around one thousand calories compared to five hundred). Americans also consume far more sugar and sweeteners than the global mean.

To find the most average eaters in the world, you need to look to some of the middle-income countries of the developing world, especially in Latin America. These countries seem to hold up a mirror to the way food consumption is now shifting to a global mean. Purely in terms of the crops consumed, one of the most average places in the world for food is Colombia. Here, the top four sources of calories used to be maize, followed by animal products, sugar, and then rice. Now the order has changed. At the top of the list of Colombian foods are animal products (518 calories) followed by sugar (404 calories), then maize (368 calories), and then rice (334 calories). Compared with the 1960s, people in Colombia have access to far more wheat and sugar and more refined oils.[32]

The idea that Colombians eat in anything like an average way would once have seemed laughable. Until recently, Colombians' food habits were not merely different from those of Europe and the United States but also distinct to the point of eccentricity from the rest of Latin America. There is nothing average about a country where people eat milk soup with eggs for breakfast, garnished with spring onions and cilantro: a dish called *changua*. To those reared on it, this soup is as soothing as congee or chicken soup. Another distinctive element of Colombian food was its unique and abundant range of tropical fruits.

On a trip to Spain in the spring of 2017, I got into conversation with the best-selling Colombian writer Héctor Abad (author of the magical and strange book *Recipes for Sad Women*). We strolled through the city of San Sebastián just before sunset, and Abad told me of his love of old books and old ways. He told me he remembered that when he first traveled from Colombia to Italy, he was astonished to find that Italians ate fruit at the end of the meal rather than at the start. In the Colombia of Abad's youth, local fruit was the opening of every dinner for those who could afford it. The fruits of Colombia range from succulent pink guavas to guanábanas, which Abad later described to me in an email as "a fruit with the peel of a dinosaur, and the meat a sweet humid cotton that you can easily chew."

When Abad was eight years old, in the 1960s, an American student named Keith came to visit his family. Keith "almost vomited" when Héctor's mother offered him *changua* soup for breakfast. Keith was also no fan of *arepas*, the Colombian corn bread that used to be ground, roasted, and baked fresh every day. Keith complained that in the city of Medellín there was not one place to get a hamburger. Abad was a teenager before he first tasted "that strange and very caloric thing called pizza."

These days, Abad and his wife still eat the good old foods of Colombian cuisine, or as many of them as they can find. They cook a lot of soups and fish or hearty dishes of meat, rice, and vegetables. But such dishes are no longer the norm for Colombians. Abad is convinced that if Keith came back to Colombia now, he would have no problem finding foods just like the ones he ate back home in Los Angeles.

Abad has noticed that young Colombians no longer eat the way that he does and that the change has happened lightning fast—"maybe five years, maybe ten," he tells me. He sees young Colombians abandon the old corn *arepas* for breakfast in favor of Westernized sliced wheat bread. He watches as they eat hamburgers and avoid the old rice and beans. He sees them sipping not fresh fruit juices but fizzy drinks, ranging from 7 Up to Colombiana—a local drink that Abad describes as "sweeter than syrup." He feels sad that the country seems to have lost its pride in the old foods. Abad's ninety-four-year-old mother still makes *changua* for herself when she is ill, but he doesn't know anyone else who does.

What's happening in Colombia is happening in most other countries too. Children around the world are now eating weirdly similar food to each other. You wouldn't expect a child in Portugal and a child in China to consume the same after-school snack. But a study conducted from 2011 to 2013 across twelve countries based on interviews with more than seven thousand nine- to eleven-year-olds found that there were very similar patterns of eating across all twelve. In particular, those children who had an unhealthy pattern of eating tended to consume near-identical foods: packaged cookies and cereal bars, branded sweets, chocolates, and crackers.[33]

Whether the children were in Australia or India, Finland or Kenya, they knew and devoured much the same things, which had nothing to do with the traditional cuisine of their country or even whether they were rich or poor. The children ate french fries and drank fizzy drinks; they ate doughnuts, chips, cakes, and ice cream. The nine-year-old in Bangalore and his or her counterpart in Ottawa had access to the same sodas, the same breakfast cereals, and many of the same bagged savory snacks. Across all the countries, the more healthy-eating children also shared similar patterns (except for the fact that children in India drank whole milk whereas those in Finland and Portugal drank skimmed milk). Children of all countries who ate healthily ate dark, leafy vegetables, orange vegetables, beans, fish, cheese, and fruit, especially bananas.[34]

If any single food illustrates the monotony of modern global diets, it is the banana. The Cavendish banana has found its way into kitchens around the world without having a great deal to recommend it as a fruit. Those soft yellow crescents have become an emblem of our food system's lack of biodiversity. They are now not only the most popular fruit in the world but the tenth most consumed food *of any kind*.[35]

THE MYTHICAL BANANA KINGDOM OF ICELAND

The unlikeliest bananas in the world grow in Iceland, a couple of hundred miles from the Arctic Circle. Iceland is not, to put it mildly, an obvious location in which to grow tropical fruit. Winter days in this part of Scandinavia sometimes have just four hours of sunlight, and temperatures regularly drop below freezing. But near the city of Hveragerði in the south of the country, there is a lava field that produces enough geothermal heat to power greenhouses where Nordic bananas grow.[36]

Homegrown Icelandic bananas are a magical proposition, one that seems to buck the trend for increasingly global, faceless modern food. Around the turn of the millennium, rumors circulated that Iceland had become "the biggest banana republic in Europe." Others spoke of Iceland attempting to become self-sufficient in the yellow-skinned soft fruits.[37]

Sadly, the "mythical banana kingdom of Iceland" turned out to be just that, a myth. Bananas may grow in Iceland—a fact that is amazing enough in itself—but that does not mean that they can be grown on a commercial scale. Back in the 1940s, when plant scientists first discovered that bananas could be grown in Iceland, there were experiments with banana farms all over the country, but they were never profitable. The growing season for Icelandic bananas is short, with harvests lasting only from April to June. Soon, the Icelandic banana entrepreneurs gave up and donated their remaining plants to the Agricultural University at Hveragerði. You won't find a geothermal banana in any local shop because the university is a publicly funded body that is not allowed to sell anything for profit. The tiny crop of bananas produced each year—about a ton—are enjoyed as a free perk by teachers, students, and visitors.[38]

For the rest of their banana needs, Icelanders do exactly the same as people in all northern and Western countries: they buy Cavendish bananas shipped in abundance from sunnier countries by a large multinational corporation. Most of the bananas in Icelandic supermarkets—and there are plenty of them—have the blue Chiquita label depicting a glamorous woman wearing a fruit-decorated Carmen Miranda hat (Miss Chiquita). Chiquita, an American firm based in North Carolina, is one of the largest global fruit brands, operating in seventy countries, selling bananas produced in South and Central America, with a large concentration coming from Guatemala and Mexico. So, far from being a banana outlier, Iceland is in fact entirely typical in the way that it consumes bananas.

Banana bread is currently one of the most eaten cakes in Reykjavik, and modern Icelanders are also enthusiastic consumers of raw bananas eaten out of the hand to gain a quick boost of energy. By 2000, according to FAO data, Iceland imported 27.5 pounds of bananas per head, nearly four times as many as Russia.[39]

Bananas are a quintessential modern food in that they are overwhelmingly grown in tropical regions but eaten in temperate ones. Bananas are grown in developing countries for the pleasure and nutrition of developed nations. Our dependence on bananas reflects the astonishing fact that it has become more common to eat foods grown from foreign crops than those from our own countries.

Those yellow fruits have gone from being something rare and specific to certain places to an ordinary presence in kitchens across the world, a foreign taste that is no longer foreign. To our grandparents, unless they lived somewhere tropical, the banana was exotic, a huge and unusual treat. Now, there's nothing unusual or exotic about bananas, which tend to be the cheapest fruit in the supermarket.

Bananas have become an everyday food in Italy and Oman, in Germany and India. Wherever in the world you eat a banana, it is likely to be the same bland Cavendish variety that dominates the world export trade, even though they never taste very good. Cavendish account for 47 percent of all bananas grown (and close to 100 percent of all bananas eaten in China and the UK).

For a long time, I was puzzled by bananas. Sometimes British people of the wartime generation would speak of how desperately they missed bananas during the war and how they yearned to eat these special fruits again when the war was over. I couldn't fathom this because the Cavendish banana is nothing to crave. But the bananas of the wartime generation were different. Before the Cavendish, the dominant banana was the Gros Michel, which was said to taste much better. It was a rare example of an old fruit that was sweeter than modern produce—and not just sweeter, but creamier in texture, with a deep, winey, and complex flavor. If you've ever eaten a banana-flavored sweet—that deep, sweetly pungent aroma—it's apparently much closer to the Gros Michel than to the Cavendish. The problem was that the Gros Michel was wiped out by Panama disease in the 1950s.[40]

When casting around for a new strain of bananas that consumers would accept, the United Fruit Company, the American-owned company that controlled most of the world's banana plantations, alighted on the Cavendish. It tasted nothing like the Gros Michel—growers at United Fruit noticed that the flavor was off and the texture was dry—but it looked the same, it transported easily, and, crucially, it was resistant to Panama disease. Without having much to recommend it in terms of flavor or texture, the Cavendish became the banana to conquer the world, largely because it looked the way people expected a banana to look. (At the time of writing, the Cavendish has been hit by a new strain of Panama disease, which casts yet more doubt on the wisdom of the banana industry investing so heavily in just one cultivar.)[41]

As a fruit engineered to be seedless, every Cavendish banana you buy is an exact genetic clone of every other banana. Bananas are the monoculture of all monocultures. There are more than a hundred varieties of bananas in existence—including red-skinned ones—but you wouldn't know it from the selection on offer in most shops, where bananas come in just one variety. Except for plantain-eaters who eat them in cooked form, you seldom hear anyone talk about the virtues of different varieties of banana because the whole point is that you expect them to taste the same: not the most delicious thing you ever ate, but cheap, filling, and fairly wholesome—compared to a bar of chocolate if not to other fruit. Bananas in the supermarket are mainly marketed not on variety or flavor but on size: small, child-sized bananas and larger ones for the rest of us.

What is true of bananas is true to a lesser extent of other fruits. There are around six thousand British heritage apple varieties, ranging from tart to sweet, from soft to hard, from yellow and green to red. Yet commercial apple production in the UK now centers on just ten varieties, chosen for their reliable look and shape and a certain bland sweetness. This varietal simplification has consequences for our health. Different apple cultivars contain varying levels of phytochemicals: vitamins that have been linked to the prevention of certain types of cancer and cardiovascular disease. If we only ever eat one type of apple, we may not get the full health benefits of eating the fruit.[42]

At least with apples, there is still a folk memory of diversity, of the enchanting old varieties that we have lost. This memory is kept alive by farmer's markets in the autumn. But with bananas, we don't even expect variety. The Cavendish is an archetypal modern food commodity. Whatever the season, it arrives hygienically zipped in its own biodegradable yellow packaging, and it has desirable, healthy overtones. Assuming you get one at the right stage of ripeness, the flavor is as consistent as Coca-Cola. You will find them in the hot summers of Dubai and the freezing winters of Iceland.

Not so long ago, Iceland was a place where fresh fruit was rare. There was a time in the 1930s when Icelanders needed a doctor's prescription to buy fresh fruit. You can see why Icelandic bananas seemed such a wonderful project to plant biologists of the 1940s. During the early twentieth century, fruit was available to Icelanders only in the summer, when there were just three kinds of native berries: regular bilberries (*Vaccinium myrtillus*), bog bilberries (*Vaccinium uliginosum*), and crowberries (*kroekiber*), a type of small black berry growing on sprawling shrubs. The crowberry is mouth-puckeringly sour. Icelandic food writer Nanna Rögnvaldardóttir notes that it would not be considered a delicacy in any country that had access to sweeter-tasting berries.

Crowberries used to be one of the inimitable tastes of Iceland, along with moss, seaweed, smoked offal, soured milk (*skyr*), and salt cod. For centuries, the people on this inhospitable island ate a diet unlike anywhere else in the world, determined by what was available. Grain was nearly impossible to grow, and so, instead of a slice of bread, Icelanders sometimes ate dried fish spread with butter.[43]

In a world before bananas (and all the changes that came with them), Icelanders were people who could appreciate tiny differences in the limited range of ingredients that they ate. In the old days in Iceland, people ate so much cod that they became intensely attuned to the variety of flavors and textures within a single fish, from cheeks to eyeballs. There are 109 words in Icelandic to describe the muscles in a cod's head.[44]

The culture that gave rise to this varied language of food has largely gone. Much of the food of Iceland is now the food of everywhere. Rögnvaldardóttir remembers a time when salt and pepper—and possibly cinnamon, for cakes—were the only spices you could find in Reykjavik. Now, Iceland—despite its cold climate—enjoys extra-virgin olive oil, sun-dried tomatoes, and garlic in profusion.

Since the 1960s, an ever-increasing range of fruits have been imported into Iceland, and there is no need to forage for sour crowberries unless you desperately want to for old times' sake. A typical Icelander today gets 109 calories a day from fruit, compared with just 46 calories in 1960. At the publishing house where Rögnvaldardóttir now works, a consignment of fresh fruit is delivered to the office every day, about half of it bananas. But in all this variety, she can't help feeling that something has been lost. "As virtually all our fruit is imported, we are rather ignorant of the seasons," she comments. Bananas are regarded with affection, she says, because they are relatively affordable and always there in the shops, even in winter. In some fundamental way, Icelanders do not know these new foods as intimately as they once knew cod and crowberries. The average Icelander eats 111 bananas every year. Yet to describe this endless feast of bananas, an Icelander has just one bland word: *banani*.

I can't entirely lament the existence of the Cavendish banana, not least because I always have them in my kitchen, ready to feed a hungry child or to slice onto morning porridge. Without the Cavendish, millions of poorer consumers would have little or no fresh fruit in their diets at all. They are a useful source of potassium, fiber, and vitamin B6. But this monoculture of fruit is a symptom of our food culture's wider obsession with cheapness and abundance over flavor. The salient fact about bananas—one of the most wasted foods in the typical home kitchen—is that there always seem to be too many of them to eat up before they turn brown.

A SHORT HISTORY OF EATING TOO MUCH

The immense volume of food in our lives is no fluke; it was planned for. In more ways than one, our food system goes back to the aftermath of the Second World War, when governments around the world became obsessed with making sure that their citizens had enough to eat, after the misery of war. In Europe and the United States, farmers were paid subsidies for the sheer volume of food that they could produce. We are still living with this legacy of quantity over quality.

Before the war, most farmers had run small mixed farms based on the principle of crop rotation to maintain soil fertility and control pests. After the war, farmers started to specialize in order to get the maximum yield possible from the land. Nitrogen was diverted from the old bomb factories to make fertilizer, and tanks were repurposed as combine harvesters. Under the US Marshall Plan, which ran from 1947 to 1952 to help with postwar reconstruction in Europe, $13 billion was pumped into the economies of the Continent. Much of it arrived in the form of animal feed or fertilizers. The era of plenty was beginning.[45]

One of the paradoxes of the postwar food system was that it entailed the greatest expansion of agriculture the world had ever seen, even as there was a mass exodus of farmers from the land. By 1985, just 3 percent of the American population was farmers, where a hundred years earlier, farmers had been more than half of the population. But the new farms did not need so many farmers, thanks to huge efficiencies of machinery and fertilizer. Between 1950 and 1990, world output of wheat, corn, and cereals more than tripled, with the United States leading the way. Something had to be done with all this grain. Increasingly, it was fed to animals to fuel a rising meat market.[46]

In this revolution of the land, we lost thousands of small farms. But what we gained was a colossal supply of calories, which after all was exactly what governments had been hoping and planning for after the war. The calories available to the average American increased from 3,100 per day in 1950 to around 3,900 by the year 2000—around twice as much daily energy as most people need, depending on their activity levels. Put another way, to *avoid* overeating in today's food environment, most of us would need to reject half of our allotted calories. Every day. This is not impossible, nor is it easy, given that it is human nature to eat whatever's available.[47]

These changes went along with the increasing dominance of huge multinational food companies who found a way to take the surplus calories and add value to them—which meant adding margins. The power accrued by these companies in the decades since the war is hard to overemphasize. By 2012, the revenue of Nestlé alone was $100 billion, twice as much as the GDP of Uganda (at $51 billion). It was these companies, more than the farmers themselves, who profited from the overproduction of subsidized crops in the West. If you break down the US food dollar now, only 10.5 percent goes to farmers. A much bigger share (15.5 percent) goes to food processing. By itself, the actual raw cereal in a box of cereal is almost worthless. What adds the value are the flavorings, sweeteners, and crisping agents, as well as the pictures on the box and the advertisements that make a child clamor for his or her parents to buy that cereal brand.[48]

Figure 1.1. Average energy use versus average energy need.

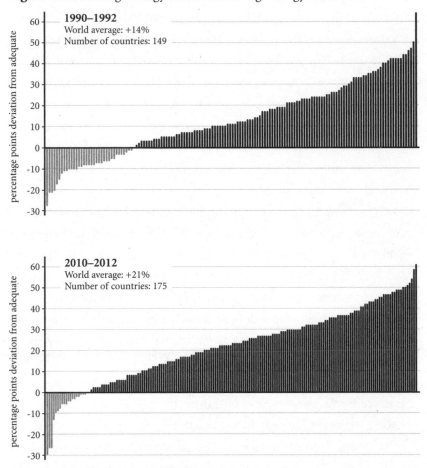

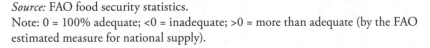

Source: FAO food security statistics.
Note: 0 = 100% adequate; <0 = inadequate; >0 = more than adequate (by the FAO estimated measure for national supply).

In the early 1990s, European governments were still subsidizing farmers to churn out mountains of food, surpluses of which often found their way onto the world market, where they made it hard for producers from poorer countries to compete. In 1995, the World Trade Organization was founded. Its aim was to end the unfair subsidies and remove trade restrictions, to give the developing world more of a level playing field. But the new liberalized global markets were not necessarily any fairer than the system that came before, and they certainly did not result in better diets. The richer countries carried on subsidizing their own local farmers but also benefited from relaxed subsidies overseas, enabling their farmers to enter new markets in the developing world. Meanwhile, rules on investing in the food markets of poorer countries were radically liberalized, which led to a huge wave of foreign investment from companies selling highly processed foods. This paved the way for the nutrition transition to happen in Asia and South America.[49]

Western eaters have been living in the sugary abundance of stage four for decades. The difference now is that so many other poorer countries are galloping to join us. In wealthy countries, the key decades of dietary change were the 1960s and 1970s, when people shifted en masse to diets higher in sugary drinks and highly processed foods. As far back as 1980, the average Canadian was already getting more than a thousand calories a day from animal products, chiefly meat, and more than three hundred calories each from oils and sugars. The great food revolution of our times is that people across the entire globe are starting to eat this type of oil-heavy, ultra-processed diet.[50]

One of the frightening things about stage four has been how fast it has happened. It took thousands of years to get from a hunter-gatherer society to one based on farming (from stage one to stage two). The effects of the Industrial Revolution in Europe and the United States took only a couple of centuries (stage two to stage three). But the new shifts in the West away from home-cooked meals and tap water and on to packaged snacks and sugary drinks were speedier still, taking only a couple of decades. In Brazil, Mexico, China, and India, the change is happening even faster, in the space of ten years or less. For South America, the peak decade of nutritional change was the 1990s. Over just eleven years, from 1988 to 1999, the number of overweight and obese people in Mexico nearly doubled, from 33.4 percent of the population to 59.6 percent.[51]

Figure 1.2.   A changing plate of food in China and Egypt in 1961 and 2009.

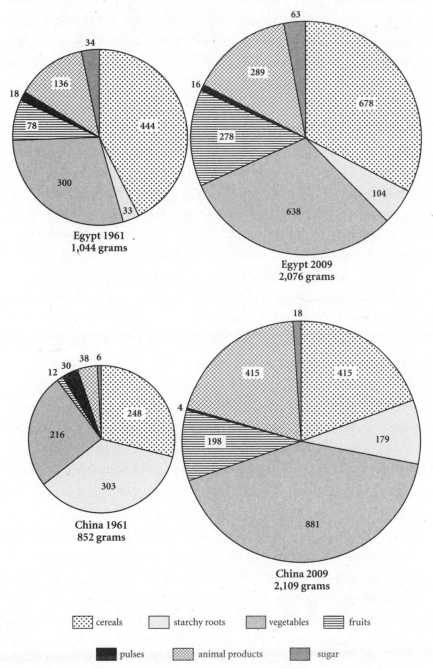

Egypt 1961
1,044 grams

Egypt 2009
2,076 grams

China 1961
852 grams

China 2009
2,109 grams

cereals   starchy roots   vegetables   fruits

pulses   animal products   sugar

Source: Sharada Keats and Steve Wiggins, 2014, "Future Diets: Implications for Agriculture and Food Prices," January, Overseas Development Institute, London.

Mexican diets have changed at tumultuous speed. After the North American Free Trade Agreement (NAFTA) was signed by the United States, Mexico, and Canada in 1994, it spelled the end of subsidies for homegrown Mexican corn, and the Mexican market was flooded with cheap yellow corn from the United States, which did not have the same qualities as the old corn, either in taste or nutrition. Traditional Mexican tortillas were made from locally adapted landrace corns of diverse species, each of which had its own distinct flavor and nutritional properties. Before it was cooked, the corn was "nixtamalized": soaked in an alkaline solution that increases the nutritional properties of the grain. The old tortillas were eaten with beans, a culinary arrangement that also reflected agricultural practice. Traditionally, in Mexico, corn and beans were intercropped to enrich the soil. Now, corn and beans are not necessarily seen together either in the soil or on the plate in Mexico. Refried beans have been edged out by ultra-processed foods, whose sales expanded at a rate of 5–10 percent a year from 1995 to 2003.[52]

As in South Africa, the pattern of eating in Mexico has changed, radically and fast. We are not talking here about the occasional fizzy drink or a Friday-night plate of fried chicken but a near total transformation of the food supply, which has gone hand in hand with disastrous changes to the population's health since the 1990s. From 1999 to 2004, 7-Eleven doubled its number of stores in Mexico. There are Mexican towns where running water is sporadic and Coca-Cola is more readily available than bottled water. Meanwhile, the prevalence of overweight and obesity among people in Mexico rose 78 percent from 1988 to 1998, and by 2006, more than 8 percent of Mexicans were suffering from type 2 diabetes.[53]

A similarly tragic nutrition transition is currently playing out in Brazil, where much of the population is both malnourished and obese. Throughout Brazil there are dual-burden households, where some of the family members are underweight and stunted (usually the children) and others are obese (usually the mothers).[54] Many adolescent girls in Brazil are both anemic and obese, suggesting that their diets, though plentiful, are low in crucial micronutrients, especially iron.[55]

To Americans, junk food is nothing new. Cracker Jack, a sticky, packaged confection made from popcorn, syrup, and peanuts, was first sold in Chicago in 1896. The difference now is the sheer global reach of branded processed items, which have succeeded in traveling to some of the remotest villages in Africa and South America. From the late 1990s onward, the multinational food companies worked hard to get their products into even the tiniest village food stores in Africa. As soon as electricity reached a given region, Coca-Cola would be there, offering free coolers and kiosks to shopkeepers who would stock their products. But now, food companies have taken this marketing a stage further, employing traveling salespeople to bring branded processed foods right into individual homes.[56]

In 2017, reporters from the *New York Times* followed some of the women who act as door-to-door salespeople for Nestlé in Brazil. Items such as chocolate pudding, sugary yogurts, and heavily processed cereals are sold door-to-door to consumers who may believe they are doing the best for their families by buying these products, which often boast that they are fortified with vitamins and minerals. In the poorer regions of Brazil, door-to-door sales enable multinational food companies to reach households they could never otherwise have penetrated.[57]

A report on the Nestlé website in 2012 boasted about door-to-door sales as a form of "community engagement" because the foods being sold were fortified with vitamins. What the report doesn't mention is that these fortified foods contain excessive amounts of sugar and refined starch and are displacing other, more nourishing foods from the Brazilian diet. The company at the time had seven thousand saleswomen going door-to-door in Brazil but aimed to expand this number to ten thousand. Nestlé claimed that the initiative brought a sense of independence as well as valuable income to the female sellers. Needless to relate, the company said nothing about the fact that most of these women—like their customers—were now grappling with diet-related ill health. A reporter for the *New York Times* spoke to Celene da Silva, a twenty-nine-year-old seller for Nestlé, who weighs more than two hundred pounds and has high blood pressure. She drinks Coca-Cola at every meal.[58]

This is a story not just about the food industry but about social change. The rise of the big multinational food companies in Brazil and elsewhere is part of a bigger picture that includes rising incomes, changing patterns of work, urbanization, electronic kitchen gadgets, and a growth in TV, computer, and mobile phone ownership. Mass media is one of the drivers of the nutrition transition that we often forget about. In China in 1989, only 63 percent of households owned a TV set, and of these, half owned a black-and-white set. By 2006, 98 percent of Chinese households owned a TV, and almost all of them showed a full-color picture. TV watching not only encourages people to be less active than ever before, but it also allows direct marketing of novel processed foods, particularly to children. Almost all of the food advertised in the world on TV is what nutritionists call "non-core": inessential sugary or salty snack foods rather than those served for a main meal. The aim of the ads is to create a preference for these unhealthy foods in children that the manufacturers hope will last a lifetime.[59]

It's not that electronic entertainment or labor-saving kitchens and city life are bad in themselves—to the contrary. Speaking for myself, I would hate to go back to a life before Spotify—never mind before refrigeration and color TV. So many of the social changes that have gone along with the nutrition transition have enabled people to lead fuller, easier, more comfortable lives. In the spring of 2018, when I visited Nanjing, one of the biggest cities in China, I walked on ground that would have been farmland ten years ago, full of workers doing back-breaking labor in the fields. Now, these neighborhoods are full of glitzy high-rise malls where young people whose grandparents had lived their youth in toil and hunger sat in air-conditioned branches of

Starbucks nibbling fluffy cakes flavored with green matcha tea. Older Nanjing residents who would once have struggled to afford exotic fruit such as durian or lychees more than once or twice a year could now buy these fruits every week and carry them home on a superfast metro train.

In a way, the modern global food industry is a miraculous achievement. It can grow anything, transport anything, and sell anything (so long as that anything can be neatly packaged and placed on a supermarket shelf). The system can produce fresh green beans and perishable meat in a far-flung corner of one country and distribute them, still in an apparently fresh state, to hungry eaters anywhere on the planet in a matter of days. For those with the cash to buy them, there are summer fruits in winter and sweet cups of piping hot chocolate topped with whipped cream all the year round. Where our ancestors worried intensely about the safety of dairy products, we can now buy fresh, refrigerated cold milk, mostly free of pathogens, whenever we feel like it.

The food transformations of stage four are unlike anything else the world has seen. Sometimes, I look at my three children and think how extraordinarily fortunate much of this generation is, never having to doubt whether there are things to eat in the shops. Fresh fruit is almost like running water to them. When stuff in our refrigerator runs out, we know there is plenty more where it came from. My children have never known empty shelves or rationing. Nor have I, come to that.

Needless to say, this diet of abundance is still not something that all children, everywhere, can rely on, even now. The terrible food shortages in present-day Venezuela are a bleak reminder that we cannot necessarily count on this era of plenty lasting forever. It's also the case that many children, even in rich countries, are not sharing in the plenty to the same extent, with one in five American children suffering from food insecurity. Stage four has seen the emergence of new forms of social and economic inequality around food. Some children have never tasted a strawberry, except for the fake strawberry flavor in a fast-food milkshake. Others—from wealthier families—are given organic oats and farmer's market berries for breakfast. The gap in quality between the diet of the poorest and that of the richest is wide and widening. The poorest families in America may not look hungry in the way that Victorian orphans looked hungry, but they eat fewer dark green vegetables, fewer whole grains, and fewer nuts.[60]

The great question held out by stage four is whether it is possible to enjoy the pluses of modern eating without the minuses. The postwar food system succeeded in delivering a vast surplus of calories. What it has not delivered thus far—at least not in most countries—is food for the masses that won't make people unwell.

BENDING THE CURVE

Experts in development studies talk about "bending the curve" of the nutrition transition, meaning changing its direction to a healthier pattern of eating. In an ideal world, we would be able to enjoy the convenience, variety, and pleasure of modern food without the chronic illness that so often seems to be its corollary. Can the curve be bent away from junk food and toward vegetables? If so, where has this ever happened?

One country—South Korea—comes up again and again when we consider these questions. South Korea managed to pass from phase three to phase four of the nutrition transition at lightning speed without experiencing anything like the same consequences of a changing diet seen in Brazil, Mexico, and South Africa. Almost alone in the world, South Korea bent the curve.

From the early 1960s to the mid-1990s, the South Korean economy was completely transformed. Between 1962 and 1996, per capita GDP increased an astonishing seventeenfold. Meanwhile, life expectancy had increased from 52.4 for Korean males in 1960 to 82.16 in 2015. As elsewhere, this growing wealth went along with demographic changes, as populations rapidly moved away from the countryside and into cities. South Koreans acquired TVs and microwaves and electric rice cookers. In 1988, the city hosted the Olympic Games and became exposed to international influences as never before.

As we might expect, these economic and social changes went along with huge adjustments to the South Korean diet. Household food consumption surveys suggest that Korean meat eating increased tenfold between 1969 and 1995—not exactly a trivial change. Previously, a dish such as spicy *bulgogi*—made from shredded beef marinated in soy sauce and sesame oil—might have been a special meal. Now, with rising incomes and falling food prices, it was an everyday midweek supper for middle-class families. Meanwhile, the consumption of cereals (centering on rice) plummeted from twenty ounces per person in 1969 to eleven ounces in 1995.

Given how rapidly South Korea moved from poverty to wealth and became exposed to new world markets, you would expect the country to have moved equally rapidly to an obesogenic diet high in sugar, new fats, and packaged foods. But compared to people in other rapidly developing nations, Koreans retained their traditional diet to a much greater extent. When researchers examined the data for South Korea from the 1960s to the 1990s, they were startled to find that South Korean fat consumption was still relatively low. In 1996, the typical South Korean ate less fat than the average Chinese person, even though the GNP of South Korea was at that time fourteen times higher than China's.[61] Meanwhile, levels of obesity in South Korea were also markedly lower than would have been expected from the nation's level of economic development. As of 1998, just 1.7 percent of men and 3 percent of women in South Korea were obese.[62]

The area where South Korea bends the curve most of all is in vegetable consumption. In 1969, the average South Korean ate 271 grams of vegetables, fresh and processed, every day. In 1995, despite all the other changes to Korean society—from the strange vogue for bubble tea to the invention of K-pop, a fusion of Western and Asian pop music—the amount of vegetables Koreans ate had actually gone up slightly, to 286 grams. The city-dwelling Koreans of the 1990s led completely different lives from Korean villagers in the 1950s, yet they continued to eat their greens. The example of Korea shows that it is possible to be a modern person who is not disgusted by cabbage.[63]

How did South Korea manage to retain its vegetable-eating ways despite all the other transformations and pressures of modern life? Part of the explanation is cultural. South Koreans see vegetables as something delicious rather than merely healthy, as we all too often see them in the West. Koreans enjoy a greater variety of flavorsome vegetables than most eaters in other countries, from bean sprouts to spinach. In rural Korea, it has been estimated that as many as three hundred different vegetables are eaten, each prized for their distinctive flavors and textures. King of all vegetable dishes in Korea is kimchi, a kind of fermented and highly spiced cabbage, which is not just a condiment but a staple food, the most consumed single item in the diet after rice, as of 2002.[64]

If Korea was helped by its vegetable-loving food culture, it also benefited from a range of government initiatives that consciously set out to soften the blow of the nutrition transition. In contrast to other developing countries, South Korea made a more concerted effort to protect its own cooking against the new globalized diet. From the 1980s onward, the Rural Living Science Institute trained thousands of workers to provide free cooking workshops educating families in how to make traditional dishes such as steamed rice, fermented soya bean foods, and kimchi.[65] In addition, there were mass media campaigns to promote local foods, with TV programs emphasizing the higher quality of local food and the benefits of supporting home-grown produce and local farmers. When most children in the 1980s switched on the TV, they would be greeted by ads for sweets and treats, fizzy drinks, and cereals. When Korean children watched TV, they might instead be fed with government-endorsed messages on the benefits of locally grown food.[66]

Fast-forward to the present day, and the average South Korean diet is no longer quite as healthy as it was in the 1990s. When Popkin returned to look again at the data on Korean diets in 2009, he found that consumption of both alcohol and soft drinks were on the rise. From the late 1990s to 2009, the Korean government put a decade of effort into promoting the consumption of whole grains, yet the average person only ate around sixteen calories more of whole grains than before. The message this time was less effective. The prevalence of obesity, diabetes, and heart disease in South Korea were also much higher in 2009 than ten years earlier.[67]

Yet the Korean diet has not deteriorated dramatically. Vegetable intake remains remarkably high, and kimchi is as popular as it has ever been. This is all the more remarkable considering that the price of cabbage—the main ingredient in kimchi—rose by 60 percent between the 1970s and 2009. The average South Korean diet might

Figure 1.3. Vegetable consumption in Republic of Korea, 1969–2009, grams per person per day.

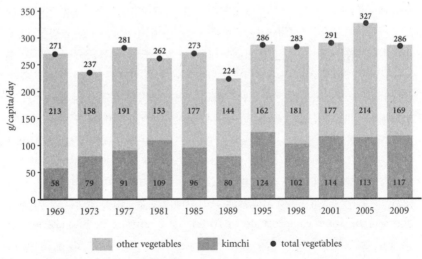

Source: Figure 2 in Lee et al. 2012.
Note: Values are presented as three-year or four-year moving averages. Kimchi intake is presented as a mean from age >= 1 between 1969 and 1995, and from age >= 2 from 1998 to 2002.

not be perfect—what human diet ever has been?—but South Korea remains a remarkable proof that it is possible to attain some kind of golden midway point between the wholesome but too-scarce diets of the past and the plentiful but unhealthy diets of the present.[68]

What South Korea shows is that the curve of the nutrition transition *can* be bent, at least slightly, with the right interventions from government. This offers hope to the developing countries of sub-Saharan Africa. Perhaps they, too, will be able to retain the best aspects of their vegetable-centric and varied diets while moving to lives of greater comfort, wealth, and ease.

At the time of writing, however, it is doubtful whether the governments of other developing countries will follow the South Korean route in actively trying to fight the onslaught of packaged foods. The more common approach seems to be for governments not so much to fight the nutrition transition but to try to make it curve faster, to gain from the profits of multinational food companies.

In August 2017, I was in Copenhagen for the World Food Summit, a two-day conference aimed at finding better ways for the world to feed itself. One of the speakers was Harsimrat Kaur Badal, the minister for processed foods in India's government (I never knew such a job existed). Badal stood up and gave a passionate speech lamenting the Indian attachment to freshly cooked food made from fresh vegetables. India, she said, was a country where most people still ate three home-cooked meals a day. The audience of Danish and international food writers, chefs, and representatives from the food industry let out a mild sigh of envy. Oh, for the fresh-cooked food of India! But the minister was trying to explain to us that this fresh and delicious food was actually a very bad and wasteful thing. "We only process 10 percent of the food we produce in India," she lamented. She compared this to the countries of Western Europe, where around 60 percent of food is processed. The minister made the point—quite reasonably— that the middle classes in India wanted to eat the same foods that were available to people with money in the rest of the world. She also pointed out that India wastes $40 billion worth of food every year, mostly because of inefficient distribution networks. "Food waste is morally wrong." The answer, she suggested, was foreign direct investment (FDI) in processed food.

India was a gigantic business opportunity, the minister explained. It was a market of 1.4 billion people, whose potential as consumers of processed food was still largely untapped. "I invite you all to come and partner with my country," she announced. "We want you to teach us your Danish technology and knowhow." In return, she offered India's amazing ingredients and a "platform" of customers ripe for the picking.

Is this really the route that the governments of India and other developing countries want to take through the nutrition transition? India is a country with a long-standing love of vegetables, which has the potential to experience something like the South Korean version of phase four rather than the health-destroying version seen almost everywhere else. Rising incomes in India are a wonderful and life-changing thing, on so many levels. But already, as India welcomes more ultra-processed food into its diet, the country is seeing an alarming rise in type 2 diabetes and insulin resistance. Is there a way for India to enjoy life beyond hunger without having to suffer the diseases of affluence?

WHAT WE ATE NEXT

Based on everything we know about history, stage four will not be the final phase of the nutrition transition, but no one can say for sure what future diets will look like. One thing that seems certain is that after fifty years or more of overconsumption, there will have to be some kind of shrinking back in the amount of calories populations consume. What remains to be seen is whether this reduction will be forced on us by climate change and failing harvests or whether we can take control of our own food destiny and start to eat within the limits of what our bodies need and what the land can bear. Barry Popkin is among those who predict that with the right policies, the latter can happen, and we will leave behind stage four for stage five, a phase of life that he has christened "behavioral change."

Stage five—if it ever fully comes into being—is where the hope lies. During this phase, most people would still be affluent and live in cities, but the cities would take on different characteristics, with more opportunities for physical exercise and more accessible and affordable fresh produce. This stage would be characterized by people eating more vegetables and fruits and experiencing a rapid decline in degenerative diseases. During this phase, greater knowledge of the links between diet and health would lead people to eat better diets. Phase five is where we would all like to be living and eating: a comfortable life with neither hunger nor disease, with delicious food but not an excess of it.

There are little glimmers that stage five may be emerging—not everywhere and for all people, but in enough places that it starts to look something like the future. One of those places is Denmark.

"So much has happened in twenty years. It is unbelievable how exciting it is to be a cook right now!" exclaims Trine Hahnemann, a caterer and cookbook author based in Copenhagen. I meet Hahnemann at the same World Food Summit in the summer of 2017 where I hear the Indian minister for processed food speak. Hahnemann takes me to a wine bar in one of Copenhagen's many beautiful old townhouses where we drink Grüner Veltliner white wine from elegant Scandinavian glasses with long stems and flattened bulbs. She tells me how she sees good food as something central to the quality of life in general.

As a Dane, Hahnemann's experience of modern food is completely different from that of her middle-class equivalent in Mumbai or Delhi. Denmark passed from stage three to stage four of the nutrition transition in the 1950s and '60s. Now, it is heading somewhere altogether more flavorsome and interesting. If stage five exists anywhere, it is surely in Copenhagen, where the majority of adults cycle to work and the food culture centers on dishes that are healthy, sustainable, and delicious. As in South Korea, Denmark benefits from a government that takes the quality of its citizens' diet seriously. In 2004, Denmark placed a blanket ban on trans fats in foods for sale, a move that played a part in reducing the country's rates of heart disease.[69]

When Trine Hahnemann was a child, no one in her Copenhagen school had heard of garlic. She remembers how long it took for hummus to be accepted by conservative Danish taste buds. "Yet now," she remarks, "you couldn't go to any food store and not find hummus. That's in thirty years. That's diversity." A mere ten years ago, there was no Vietnamese food to be found in Copenhagen, whereas now there is a passion for pho, a spicy Vietnamese broth heady with green herbs and vegetables. Yet the Danes have also retained their love of healthy traditional foods such as dense, dark rye bread.

As someone who caters for a government-funded work cafeteria, Hahnemann has seen firsthand how the Danish government makes healthy and sustainable eating a priority for everyone in society, rich and poor. As of 2016, a new law came in requiring that any food served in a public institution—from a school to a hospital—must be 60 percent organic. Hahnemann finds that the Danes she cooks for are remarkably receptive to vegetables and flavorings that would once have been seen as threatening. If a big batch of cauliflower arrives from her suppliers, she may serve it three days in a row: day one with a brown butter sauce, day two as Indian pakoras, and day three Italian-style with capers.

Despite Hahnemann's love of vegetables, not everything she cooks and eats would be defined by a nutritionist as strictly healthy. Like most Danes, she adores cake and always keeps a sponge cake in her freezer in case friends drop by and she wants to rustle up a quick rhubarb and chocolate layer cake, filled with a rich rhubarb cream and topped with chocolate ganache. "Life without cakes would be a bit too sinister," Hahnemann writes in one of her cookbooks, adding that she believes cake to be good for mental health. Just as the Japanese-style diet eaten by Fumiaki Imamura is a mix of healthy and unhealthy, so is the modern Danish diet. But the balance, in both cases, is tipped toward the healthy.[70]

Not every country can be like Denmark, which benefits from a tiny population, substantial wealth, and low levels of social inequality. It would be difficult to replicate exactly the way that the Danes eat anywhere else. The question, however, is whether other countries could shift to a phase where the typical diet is something abundant that no longer damages the health of millions.

There are a few small but growing signs that many people around the world are moving in something like a Danish direction with food. Fumiaki Imamura's data shows that the quantity of healthy food being eaten in the more affluent countries of the world—in Europe, North America, and Australasia—is actually going up, while consumption of unhealthy foods slowly starts to level off. This aligns with the behavior about food we can observe all around us, with many consumers consciously reacting against what they see as a toxic food supply and searching for new ways of eating. Who would ever have predicted the day when kale and beetroot would become objects of affection in the West? Food preferences can change in a remarkably short space of time.

The hope held out by stage five is that the two stories of modern food could merge into one, single story: a cheerier and more consistent one. We abolish hunger, eat our greens, make water the default drink, discover delicious things like hummus, have the occasional slice of cake for our mental health, and live happily ever after. With the right food policies—which would include a combination of different farming policies, better food education, and tighter regulation of unhealthy foods and drinks—we might yet reach stage five. For this to happen, governments would have to reset the trajectory of food policy away from the postwar agenda of quantity at all costs. There are tiny signs of this happening—for example, in the sugar taxes that have been enacted in various countries—but the true potential of food policy to improve our diet has yet to be tested. As the authors of one briefing paper on future diets from 2014 remarked, "Policies on diets have been so timid to date that we simply do not know what might be achieved by a determined drive to reduce the consumption of calories, and particularly the consumption of fat, salt and sugar."[71]

In the meantime, for those of us still in the middle of stage four, it can be hard to know how to live and eat for the best. We are beset on all sides by extremes—from fad diets on the one hand to junk food on the other—and it can feel almost impossible to steer our own path through the madness and choose a variety of foods that give us both pleasure and health.

It would be a start if we could at least name the food in front of us and notice what it is that we are putting in our mouths. Half the time, we do not even seem to recognize the ways in which our food has changed.

Colin Khoury, the diversity expert who identified the Global Standard Diet, tells me about a dinner table game he plays at his home in Denver. Khoury lives with his wife and disabled brother and all three play the game every evening when they eat. It's a kind of secular grace. Before taking the first mouthful, the three of them compete to name the species and botanical family of all the foods they are about to eat. In Latin. If they are eating burritos, for example, one of the Khourys might start by saying thanks for *Triticum aestivum* (wheat, in the tortilla), in the grass family (*Poaceae*). Then another person might say, "*Persea americana*, avocado, in the laurel family, *Lauraceae*." They keep going until no one at the table can name any more ingredients. And then the three of them eat. "It's kind of a silly exercise," says Khoury, "but for me it's a chance to pull it apart enough to be recognizant of what's in a meal." Khoury's dinner table game is a small but eloquent gesture against a world converging on the same unbalanced diet.

Personally, my Latin is not good enough to play this game. But I like Khoury's idea of picking apart the components of food on a plate as a way of paying attention to what you are actually eating. This is what omnivores have always done: we look at a range of items and say "this is edible" and "this isn't." None of us can escape living and eating in the global market of stage four. You can't increase the variety of your daily food intake simply by naming it. But if we are going to tip the balance of our diets back in a better direction, it helps if you can at least say what it is that you are eating.

One of the problems with modern eating is that we stopped trusting our own senses to tell us what to eat. We may not be hunter-gatherers, or even farmers. But every human is still an eater, and we still have senses that can tell us useful things about what to put in our mouths, if only we pay attention. You are under no obligation to eat something just because a packet tells you it is "all-natural" or "protein-boosted" or supposedly marvelous in some other way. Despite all the transformations of stage four, some things remain constant in our eating lives. Food is only food when a human says it is, and that human is you.

CHAPTER TWO

MISMATCH

"SOMETIMES WE NEED TO STEP BACKWARDS." THUS begins one of the many voices on the internet suggesting that our eating would be healthier and happier if only we could travel in time and eat a bit more like our great-grandmothers. This particular article—from the Institute for the Psychology of Eating—goes on to recommend "ancestral eating" as the solution to many of the health problems of modern times.

What, you may ask, is ancestral eating? Apparently, it means sticking as closely as possible to the diet of your great-grandparents, wherever they happen to have lived. If your ancestors came from Greece, ancestral eating might entail full-fat yogurt, wild greens, grass-fed meat, and olive oil; if your family came from Japan, it might include fish, seaweed, fermented vegetables, and heirloom grains.[1]

Nostalgia for the tastes of our childhood has always been a powerful emotion. In our modern food environment, many of us invoke the wisdom of our grandparents as a way out of the craziness and ill health of modern diets. The inspiration for much of this way of thinking comes from the food writer Michael Pollan, who memorably advised that a good rule of thumb for healthy eating was "don't eat anything your great-grandmother wouldn't recognize as food."

The urge to turn back the clock on modern diets is understandable. So many aspects of our diets *have* worsened in recent decades. In all regions and in all countries, diets rich in coarse grains, legumes, and other vegetables are disappearing as a mainstream way of eating and, as we saw in Chapter 1, there has been a great loss of biodiversity. It's true that almost anyone in the modern world would be nutritionally better off eating more olive oil, more vegetables, more fish, more lentils, and more whole grains.[2]

Yet there are significant problems with thinking that the solution to poor diets is to go backward. For one thing, our great-grandmothers often suffered terribly for the food they made, as they toiled to grind enough grain to keep their families alive. Until recently, it was common for women in much of the world to suffer from severe arthritis in their upper bodies, caused by the hours they put in at the grindstone and rolling dough for such staples as chapattis and tortillas.

Moreover, not all great-grandmothers were eating an ideal diet. Many of our recent ancestors, as we've seen, were eating an extremely monotonous diet of grains and teetering on the brink of hunger. True, your great-grandmother wouldn't recognize sports drinks or popcorn fried chicken or any of the myriad other new highly processed foods, but she also might not recognize many of the wholesome new foods that contribute to health: raw kale salad and overnight oats and pumpkin seeds. Some great-grandmothers, moreover, were eating an early twentieth-century version of junk food. In 1910, a public health campaigner in New York City watched schoolchildren buying hotdogs dyed with violently pink food coloring and frosted cupcakes. It's simply not true that our great-grandmothers would only recognize meat that was organic or grass-fed.

There is yet another difficulty with calling on the wisdom of our great-grandparents to save us from the worst excesses of modern food. This way of thinking ignores the fact that we are already living and eating with one foot in the past. Many of our most profound problems with eating stem from our inability to fully adapt to the new realities of the nutrition transition. In many ways, we already *are* eating according to the wisdom of our great-grandmothers, whose physiology and attitudes to eating were forged by the constant threat of scarcity.

What we eat may have radically changed in our lifetimes, but our food culture has not changed quickly enough to keep pace. We may, sadly, have forgotten the recipes of our great-grandmothers. Most of us have lost their homespun knowledge of how to bottle fruits for the winter, not to mention their brilliance with a carving knife. But what we have not forgotten is their excitement at a laden table. We are living in a world of perpetual feast but with genes, minds, and culture that are still formed by the memory of a scarce food supply. This is part of what it means to live through the vertiginous changes of stage four. We haven't yet developed the new strategies for living that would enable us to navigate our way through this forest of seeming plenty to a way of eating that gives us both health and pleasure.

Think about some of the eating strategies that would have made sense in an era of scarce food. For one thing, you would value energy-dense foods such as meat and sugar very highly and gorge on them when they came your way—just as many of us still do. You would leave a clean plate, and when food was accessible, you would grab it while you could.[3]

Development experts speak of "mismatch" in explaining the clashes between the new food reality and the persistence of a human biology and culture adapted to earlier times. Instead of looking backward to some imagined past that we can never reclaim, we need to look forward and have yet another change of taste.

Our food system is currently full of mismatches. Some of these mismatches are cultural, as we fail to adapt to the new realities of eating in an age of abundance. Our food culture remains far too misty-eyed about sugary foods, for example. We haven't adjusted emotionally to the fact that sugar is no longer a rare and special cel-ebration food, worthy of devotion. Nor have we yet modified our attitudes toward those who are overweight and obese to reflect the fact that these people are now in the majority.

Perhaps the most tragic mismatches are biological, as bodies formed for an environment of scarcity have not adapted to cope with the strange and bountiful new world we now find ourselves eating in.

THE THIN-FAT BABY

It was 1971 and Dr. Chittaranjan Yajnik was a young medical stu-dent training at Sassoon General Hospital, Pune, a big city in the west of India. Yajnik was given the task of measuring the body mass index (BMI) of diabetic patients. This should have been a routine job, little more than number crunching. The main challenge was that Yajnik could not afford a calculator, so he laboriously wrote down the patients' weight in pounds and height in feet in a log table and used his paper notes to convert the measurements into metric ones and then calculate in his head the BMI in kilograms per meter squared.[4]

After taking measurements for the first ten patients, Yajnik noticed something was not right about his numbers. His medical textbooks had taught him that type 2 diabetes was a disease mostly suffered by the old and the obese. But the first ten diabetic patients that Yajnik measured in the hospital at Pune were all young and thin, with low body mass indexes. If his measurements were correct, then the textbook must be wrong, or at least incomplete, in its definition of type 2 diabetes as an offshoot of old age and obesity. Yajnik tried to raise the problem with his medical supervisor but was told that this was no time to be challenging medical orthodoxy—he should just focus on passing his exams.[5]

Yajnik could not put the puzzle of diabetes in India out of his mind. After some years studying Western diabetes in Oxford, Eng-land, he returned to Pune as a fully qualified medical researcher, by which point diabetes was on the rise in his home country. In the early 1990s, Yajnik began a study following mothers and their babies in six rural villages near Pune—the Pune Maternal Nutrition Study. The data he started to gather confirmed his hunch that diabetes in India had a very different face from the supposedly classic type 2 diabetes in the textbooks. Yajnik took detailed birth measurements of more than six hundred Indian babies and compared them with a cohort of white Caucasian babies born in Southampton in the United Kingdom. Compared to the UK babies, the Indian babies were smaller and lighter. Yet when Yajnik used calipers to measure the thickness of the babies' skinfolds, he found that the small Pune babies were actually fatter than the Southampton babies—they were surprisingly adipose, especially around the center of the body. Yajnik coined the phrase "the thin-fat Indian baby" to describe this phenomenon. Even at birth, these Indian babies had higher rates of prediabetes hormones in their bodies than their British equivalents. The babies may have looked thin, but their body composition was actually fat.[6]

We speak of conditions such as heart disease and type 2 diabetes as noncommunicable diseases, or NCDs. You can't catch an NCD from another person in the way that you would catch a common cold by standing next to someone who is sneezing. But what Yajnik discovered is that babies can actually "catch" a predisposition toward diabetes from their mother in the womb, via the diet she eats. The babies of mothers who were undernourished during pregnancy had "fat-preserving tendencies"—passed on as a survival mechanism.[7]

It used to be believed that India's diabetes epidemic was mainly due to "thrifty" genes, endowed over many generations on populations that suffered from patchy and inadequate food supplies. Thanks to decades of malnourishment, these populations were poorly adapted to eat a rich modern diet. Yajnik's breakthrough was to show that the time frame of maladaptation was much shorter. He speaks not of a thrifty gene but a "thrifty phenotype": the interaction of genes with the environment over a single generation. Depending on the environment in which it develops, a given gene may give rise to different phenotypes. The thin-fat baby represents a mismatch of biological environments. These babies grew inside their malnourished mothers with phenotypes for hunger but—thanks to the huge changes in India's food supply between the 1970s and the 1990s—found themselves eating an unexpectedly plentiful diet.[8]

When Yajnik first observed the thin-fat baby in the 1990s, this was a radically new way of thinking about the interaction of nutrition and health. It took six years for Yajnik to have his first paper on the subject accepted for publication because the mainstream medical establishment was so skeptical of this idea "coming from an obscure Indian in an obscure place," as he puts it. The idea of the thin-fat baby started to gain acceptance only when Yajnik published a paper in 2004 revealing that he was a thin-fat Indian himself.[9]

This 2004 paper—which he called the "The Y-Y Paradox"—included a now-famous photograph of Yajnik side by side with his friend and colleague John Yudkin, a British scientist: two slim middle-aged men in white shirts. The paper explained that Yajnik and Yudkin had near-identical body mass index readings of 22 kg/m². A BMI of anything between 18.5 and 24.9 is considered healthy in the United Kingdom: not underweight and not overweight. Yajnik and Yudkin were both well within this healthy range. But x-ray imagery showed that Yajnik—the thin-fat Indian—had more than twice the body fat percentage of his friend. Yudkin's body fat was 9.1 percent whereas Yajnik's, despite his slim appearance, was 21.2 percent. Further research has confirmed that the adult Indian population in general has lower muscle mass and higher body fat than white Caucasians or African Americans.[10]

The story of the thin-fat babies of India is the story of the nutrition transition written on human bodies. Thanks to the new science of epigenetics, we now know that a pregnant woman's body sends signals to her unborn child about the kind of food environment he or she will be born into. An underweight pregnant woman who eats a scarce diet is signaling to her child that food will always be scarce, which triggers a series of changes in the baby's body, some hormonal and some physiological. For example, Yajnik found that a lack of vitamin B12 in the mother's diet resulted in babies who were more likely to be insulin resistant.

Thin-fat babies are graphic evidence of a society in a state of dietary flux, with a shift from starvation to abundance in a generation. These Indian babies were born to mothers who lived and ate not so long ago, but the circumstances of their lives feel like another universe. There was seldom enough food, especially fats and protein, and people had to walk many miles just to get fresh water. When these women became pregnant, their babies' bodies were metabolically programmed before birth—with their ample deposits of abdominal fat—to survive in circumstances that were harsh and lean. But the babies grew up eating in a very different and more affluent environment: a world of improved buses and electricity and labor-saving farm machinery, of cheap cooking oil and rising incomes. Millions of people in Indian cities—a new and rising middle class—have scooters where once they had only bicycles or feet. Diabetes is the worm in the apple of this new Indian prosperity.

The problems of babies born into a rapidly changing food environment are compounded by the way they are fed during the early years of life. The memory of scarcity still informs the strategies mothers use to feed babies, not just in India, but everywhere in the developing world. Many of the thin-fat babies will have been fattened up in their first two years by emergency food aid. In the old India, the most urgent nutrition problem was outright hunger, and overfeeding a child seemed like the last thing anyone should worry about. This hungry India still exists to a shocking extent, with 38 percent of all children under five so short of food that it will impair their future development, according to the Global Nutrition Report. If the alternative is to starve, rapid weight gain in the first two years of a child's life can be a miracle. But it's now known that this rapid growth in children who were previously malnourished may have unintended long-term consequences. Rapid growth is a risk factor for obesity and elevated blood pressure in later childhood and diabetes in adulthood. There is gathering evidence that high intake of protein and vegetable oils during the early years of feeding may result in a higher risk of obesity later in life.[11]

Given India's vast population, it is perhaps not so surprising that the country currently has more patients with type 2 diabetes than any other country in the world. The more startling fact is that people with diabetes form such a high percentage of that population. Already, in large cities such as Chennai, around two-thirds of the adult population is either diabetic or prediabetic.[12]

What can be done to correct the nutritional mismatch suffered by the thin-fat babies? Those working with malnourished babies in developing countries have started to talk of optimal nutrition: the kind of childhood diet that will provide all the essential micronutrients and promote growth while minimizing excess weight gain. Yajnik and his colleagues are currently working on a project giving a cohort of adolescent girls vitamin supplements that should, in theory, mean that in pregnancy their bodies will send the message to their unborn children that a world of plenty awaits them. The aim of the project is to get the bodies of the mothers to communicate more accurately with their unborn children about what food is like in modern India and thus to reduce the risk to future generations of developing NCDs. Only time will tell if these hopes come to fruition. The epigenetic messages in our bodies cannot be rewritten straight away.

Spare a thought for the now grown-up thin-fat babies of the 1980s and 1990s, many of whom are now diabetics living in modern India. Through no fault of their own, these young people are stuck with a disease they will spend a lifetime trying to manage. Living with type 2 diabetes means living on a diet that is directly at odds with the prevailing food supply. In food markets awash in lavish amounts of refined carbohydrates, they must teach themselves to be sparing with sugar and white rice. They must try to limit their calorie intake in a world that offers them ever-larger portions.

The dilemmas faced by the thin-fat Indian are an extreme version of the problems facing millions of others in the modern world. We are all affected to some degree by a series of biological clashes between the basic instincts of our bodies and the environments in which we live, and taken together, these clashes seem almost designed to make us fat. Every human baby has an inbuilt preference for sweetness, which didn't matter too much in the days when sugar was a luxury, but which becomes a problem in a world of cheap sweeteners. We also have a natural inclination to conserve energy, which served us well as physically active hunter-gatherers and farmers but doesn't pan out so well in cities full of cars. Many of the human instincts that evolved to help us survive have now become a liability. Yet another example is the fact that in human biology, hunger and thirst are two separate mechanisms, something that means we can drink almost any amount of sugary drinks without deriving much satisfaction from them.

THE THIRST CONUNDRUM

Where do you draw the line between a drink and a snack? These days, it can be hard to tell. If you eat a serving of chocolate ice cream, it counts as dessert and gives you approximately two hundred calories. But if you take the same chocolate ice cream in the form of a large milkshake, the serving size may yield as much as one thousand calories. Yet because it's only a drink, you might have a burger and fries alongside.

It doesn't make sense to talk about changes to eating habits without bringing in the revolution in what we drink. Perhaps no single change to our diet has contributed more to unthinking excess energy intake than liquids, both soft and alcoholic. We have reached a state where many people—adults and children alike—can no longer recognize a simple thirst for water, because they have become so accustomed to liquids tasting of something else.[13]

By 2010, the average American consumed 450 calories a day from drinks, which was more than twice as many as in 1965: the equivalent of a whole meal in fluid form. Whether it's morning cappuccino or an evening craft beer, a green juice after a workout or an anytime bottle of Coke, the choice of calorific beverages available to us has become immense and varied. Around the world, there are bubble teas and agua frescas, cordials and energy drinks. And then there are all the newfangled craft sodas infused with green tea or hibiscus that pretend to be healthy, even though they probably contain nearly as much sugar as a Sprite. Many modern beverages are better thought of as food than drinks, judging by the number of calories they contain. Yet for reasons both cultural and biological, we don't categorize most liquids as food. To our bodies, this endless stream of drinks registers as little more satisfying than water.[14]

Picture a typical day for an average Westerner, and start counting the drinks. It's a *lot*. It surprised me to learn that more than 5 percent of Americans now start the day with a sweetened fizzy drink, but then again, cola for breakfast is a logical enough choice if you work early shifts and don't have access to a kitchen. A more universal morning drink is coffee, which is often more milk than coffee. Maybe there's an orange juice on the side. (After decades of growth, however, our appetite for orange juice is finally waning, hit by growing consumer awareness that it is little more than sugar. From 2010 to 2015, the amount of Tropicana fruit juices consumed in the United States dropped 12 percent.) By midmorning, survey data suggests that 10 percent of Americans are ready for another coffee or soda. Personally, I am in awe of anyone who waits that long. I am so addicted to coffee, particularly when working, that I am often thinking about my second cup before I have finished the first (which is one of the reasons why I try to take my coffee black as the default. Try).[15]

And so our days continue, punctuated by sips of sugar-water and caffeine of one kind or another, with or without the addition of milk and various syrups, until the cocktail hour arrives, time for more soft

drinks or alcohol. We sometimes imagine that the Mad Men generation of the 1950s were much bigger drinkers than the average person
today. But except for a small affluent minority, Americans consumed
vastly less alcohol in the 1950s than today—total alcohol intake
increased fourfold from 1965 to 2002 in the United States.[16]

This is a global story. A rise in beverage consumption is one of the
key elements in the nutrition transition, wherever it has happened.
In 2014, a market report on soft drinks wrote of Latin America as
"the global bright spot for soft drinks brand owners and bottlers."[17]
Young people in the emerging economies of Mexico and Argentina
drink more of these drinks every year, as incomes increase. In China,
people who lived their whole lives drinking nothing but unsweetened
tea and water now have access to beer and fizzy drinks and a whole
smorgasbord of Starbucks flavored coffees.

It's a sign that times are good when you can afford to quench your
thirst with something other than water. The drinks industry—both
soft and alcoholic—has conditioned us to believe that whatever the
occasion, it will be improved with a drink in our hand. Studying? An
energy drink will help you concentrate. Out with friends? You need a
beverage to help you relax. By 2004, the average American was consuming 135 gallons a year of beverages other than water—around one
and a half liters a day.[18]

It would be easy to paint all this modern beverage consumption as a novel kind of gluttony, something that those wise
great-grandmothers of ours would never have indulged in. But in
middle-income countries such as Mexico where much of the water
supply is unsafe, buying soft drinks can be a move of self-preservation.
Bottled drinks do not contain the bacteria of unclean water and are
less likely to make you and your children sick. What's more, a fizzy
drink can look like the frugal choice. Given the option between paying a similar amount for a bottle of water or a bottle of cola, the cola
can appear to be better value, because it offers flavoring and energy
along with the liquid.

But our biology is not well adapted for this switch to high-calorie beverages. When we talk about what's wrong with modern drinks, we talk a lot about the problems with sugar, but what we don't talk about so much is our own hunger and fullness. It seems that our genes have not evolved to be satisfied by drinking clear liquids, even when those liquids contain as much energy as a three-course lunch. This is the liquid conundrum. A person might easily drink two large glasses of chardonnay before dinner, then go ahead and eat a substantial meal as if nothing had happened (or maybe this is just me). Another person might have half a liter of Mountain Dew and feel no less hungry for a foot-long sandwich. With certain exceptions, our bodies simply do not register the calories from liquids in the same way that we do with solid food. This is one of the starkest mismatches between human biology and our current patterns of consumption.

Before the first experiments with honey wines around eleven thousand years ago, the only drinks available to humans were water and breast milk. For most of our evolution as a species, except for when we were infants, drinks and food were thus two entirely separate things. There were survival benefits to keeping the mechanism of thirst separate from the mechanism of hunger. If hunter-gatherers had become full from drinking water, they wouldn't have felt the need or desire to go out and search for food, and they would have died.[19]

Numerous studies have shown that most people do not compensate for the energy they drink by eating less. When you drink water, it rapidly enters your intestine, quenching your thirst but doing little to dent your hunger. The same is true even when the water is laced with sugar. It's as if our bodies simply don't register the calories in the same way when they arrive from a glass, a cup, or a can. Clear fluids such as sports drinks, fruit juices, cola, and sweetened iced teas seem to be particularly bad at killing hunger, but milk-based drinks such as lattes and chocolate milk are also surprisingly unfilling for most people, despite the nutrients they contain. Scientific studies show that people have a weak satiety response to clear drinks regardless of how many calories they contain—meaning that they don't fill us up as much as the equivalent calories taken as food. And so we end up consuming a lot more energy from drinks than we intended or even knew.[20]

As of the year 2000, sugary drinks were the single largest source of energy in American diets. Westerners have been drinking sugar-sweetened tea and coffee for a few centuries, but never before have caloric beverages taken up so large a proportion of the average diet. In the past, the largest source of energy in human diets would have been a staple food that actually filled a person up, such as bread. It's a sign of how disconnected we are from our own hungers that we have reached the point when so many people receive most of their energy from something that gives our stomachs so little satisfaction.

The relationship between liquids and hunger is still not fully understood. One biological explanation for our lack of fullness after a drink is that the normal hormones—peptides—that are triggered in our gut when we eat food are not triggered when we drink sugary or alcoholic drinks. The role of these hormones is to signal to our brains that we are full. When we have a large sugary drink, there is faulty communication between our gut and our brain, and somehow we don't get the message that we have just ingested hundreds of calories.

We need a way to think about liquid-fullness as well as food-fullness. I've found it helpful to start telling myself that anything other than water is a snack, not a drink: something to be savored rather than gulped down. A cappuccino can taste amazingly creamy and delicious when you tell yourself it's food. Whether this kind of mindful drinking would work when you have just ingested three beers and are wondering about a fourth on a Friday night is debatable, however.[21]

There are exceptions to the rule that liquids don't fill us up. After all, breast milk is both food and drink to a baby. Some liquids—soup being the prime example—are actually even more filling for most people than solid food. The thickness or viscosity of a liquid seems to be important for whether it is filling or not. The more viscous a liquid is, the more it suppresses hunger.[22] Our beliefs about different liquids may also affect how much satisfaction they bring us. Soup has a long-standing reputation of being satisfying—something that nourishes us and feeds us, body and soul. A cold, fizzy drink, by contrast, has no such nourishing connotations.

The rise of highly marketed calorie-filled drinks is a big part of why our energy balance—calories in and calories out—is so out of sync. The average BMI of the US population has been increasing for over 250 years, but it only took a sudden sharp turn upward in the mid- to late 1970s. This was the same moment when the daily energy gained from beverages suddenly increased—from 2.8 percent of all energy to 7 percent for the average person. Correlation is not causation, but the timing supports an association between rising beverage intake and rising obesity. The correlation between a sudden rise in consumption of caloric drinks and rising BMI is something that maps onto the whole population, across all ages and ethnic groups.[23]

Mainstream opinion will—charmingly—tell a person that if he or she is fat, the reason must be a lack of willpower. But the example of calorie-laden drinks shows once again that obesity cannot simply be attributed to individual laziness or greed. Around forty years ago, companies began marketing a completely new set of drinks to American and European consumers. Another couple of decades on and these novel liquids were traveling the world and becoming ever larger. In 2015, Starbucks marketed a cinnamon roll–flavored Frappuccino that contained twenty teaspoons of sugar (102 grams) in a single serving. In some ways, the surprise is not that two-thirds of the population in the United Kingdom and United States are overweight or obese but that one-third of the population are not.[24]

Yet we live in a culture that says that despite all this sugar being pumped into our drinks, we are not allowed to be fat. This is one of the cruelest aspects of our current food culture. There is a huge mismatch between the availability of food and drink and the way we talk about the people who consume the most everyday and easily available items.

THE STIGMATIZED MAJORITY

One of the most striking differences about the way we eat now compared to the past is that in most countries, the majority of people eating today are overweight or obese. Yet there has been remarkably little discussion of how the overall experience of eating is affected by this change. We wring our hands about the obesity crisis, but we do not pay much attention to how it feels to eat in the modern world as a person with obesity. Culturally, we have not yet adapted to the new reality, and we continue to hold up slim bodies as "normal." This is sad, not least because the psychology of fat shaming is one of the reasons most people with obesity find it so hard to lose weight.

The fact that weight stigma is a problem has been known since the 1960s. A series of studies done by sociologists in the early 1960s found that when shown pictures of six children and asked to rank them in order of preference, ten-year-old American girls consistently ranked a girl with obesity as the least preferred, lower in the friendship stakes than a child in a wheelchair, a child with facial disfigurement, or a child with an amputated arm.[25]

In 1968, a German American sociologist named Werner Cahnman published an article titled "The Stigma of Obesity." In it, Cahnman documented the terrible discrimination suffered by young people with obesity in America, based on a series of thirty-one interviews he conducted at an obesity clinic in New York City. They told him stories of rejection and ridicule, of doors slammed and opportunities lost. Rejection of people with obesity "is built into our culture," wrote Cahnman. In 1938, as a young Jewish man, Cahnman had been interred in the Dachau concentration camp. After his escape, he immigrated to the United States, where he spent much of his career as a sociologist considering the various forms that social prejudices took.

To him, it was clear that being overweight in America was not just seen as detrimental to health but as morally reprehensible.[26]

The worst aspect of weight stigma, Cahnman suggested, was that it created an internalized sense of shame from which the obese "cannot free themselves." In the fifty years since Cahnman's article was published, much research has been done confirming that weight stigma has damaging effects on the health and well-being of the stigmatized.

Yet stigma about being overweight or obese goes almost unchallenged. Negative messaging about fatness is the norm rather than the exception in our culture and has become a global phenomenon. There used to be numerous cultures where nonthin bodies were celebrated, but a study from 2011 found that fat stigma has now spread to Mexico, Paraguay, and American Samoa. Meanwhile in Western societies, psychologist A. Janet Tomiyama has found, weight stigma is now "more socially acceptable, severe, and in some cases more prevalent than racism, sexism and other forms of bias."[27]

Clearly, not everyone who is overweight or obese is equally sensitized to negative stereotypes about weight. Some are cheerfully unfussed by such questions as body mass index, while others find solace and pride in the body acceptance movement that celebrates human bodies in all their diversity. Nevertheless, the indications are that millions of people worldwide are negatively affected—both psychologically and physically—by obesity stigma.

The history of public health is littered with examples of health-related stigma, and it never ends well for those affected. Cholera, syphilis, and tuberculosis were all impossible to bring under control when the sufferers were seen as morally to blame. In 2017, an editorial in the British medical journal *The Lancet* argued that health systems will never effectively prevent childhood obesity until it stops being treated as a personal moral failing caused by faulty willpower. Until there is collective recognition that obesity is "not a lifestyle choice," argues *The Lancet*, the prevalence of obesity is unlikely to be reduced.[28]

In the absence of collective action, the main way anyone can shield themselves against an obesogenic world is through embarking on an individual program of diet and exercise. Yet here, again, the stigma of weight serves to thwart us. There is now a gathering body of evidence that obesity stigma negatively affects a person's efforts to lose weight. This will come as no surprise to anyone who has ever tried to force themselves on a diet only to fall rapidly off the wagon, thwarted by a debilitating sense of shame.

When I was an overweight teenager, the whole experience of eating was quite different to how it feels to me now, as a so-called normal weight middle-aged woman. It was the difference between eating in an atmosphere of freedom—as I am lucky enough to do now—and eating in a cloud of judgment. I remember feeling that I did not really have permission to eat most of the foods that I wanted to eat, especially in public. Because of my age (I was born in 1974), I feared fat more than carbohydrates. This translated into years of pointlessly and joylessly denying myself butter, because you were not allowed to eat butter—or so I believed—if you were bigger than a size 10 (a US size 6).

During my overweight years, I had two completely different ways of eating, one for public and one for private. In public—most of the time, anyway—I ate what I believed to be socially acceptable. When I was at university, my best friend was anorexic, and I felt that if I followed her lead, I must be above reproach. I ate dreary salads made from shredded iceberg lettuce and dried-up chicken breast with no dressing. I ate tiny portions of unseasoned poached salmon and cottage cheese. I drank gallons of Diet Coke. All of these items had an unpleasant overtone of duty.

Behind closed doors, the way I ate was a different story. When you have been forcing yourself to eat small pinched amounts of things you don't like all day, the urge to eat large quantities of things you do like can be overwhelming. I ate entire packets of sweet biscuits, crunchy peanut butter straight from the jar, multiple slices of toast.

I ate dinner and then had McDonald's straight afterward because the dinner hadn't satisfied me. I comfort ate as if my life depended on it. One of the things I was comforting myself for was my distress at being overweight.

I was not alone in being driven to eat in secret by a sense of shame about how I looked. Many studies have shown that experiencing weight stigma makes it more likely for someone to engage in binge eating in secret. In one study of more than 2,400 overweight and obese women, nearly 80 percent reported that they coped with weight stigma by eating more food. Other studies have shown that being teased about one's weight is linked to lower levels of participation in sports and other physical activities. When I used to go running back then—which didn't happen very often—I felt embarrassed by how I looked in my workout clothes. Each step felt like a mathematical form of atonement: so many steps to compensate for so many calories. By contrast, when I go running now, there is no particular agenda, except enjoyment, and so I am able to run for longer without worrying about anyone staring.[29]

A common misconception is that weight stigma will motivate people to lose weight. Yet making people feel awful about themselves is unlikely to spur them on to change their diets. To the contrary, we know that people who feel stigmatized by obesity are more likely to avoid healthcare settings, and who can blame them, given the judgmental way that many doctors speak about weight? There may also be more biological reasons why stigmatizing obesity entrenches weight gain. Feeling victimized is very stressful, and it is well established that cortisol, the main human stress hormone, encourages overeating. It's known from rodent studies that cortisol messes up the normal cues for hunger and fullness. High cortisol levels in the general population are very consistently associated with higher levels of abdominal fat in the body.[30]

Another mechanism through which stigma leads to weight gain is discrimination—for example, in the world of work. There is an obesity wage penalty, particularly for women. People with obesity have reported that they receive fewer promotion opportunities and less training at work than their non-obese colleagues. All of these forms of economic discrimination make it still more difficult for an obese person to lose weight. A lower income can translate into more limited food choices as well as the necessity to live in a neighborhood with crowded housing, poor access to healthy food, and few safe places in which to walk or jog.[31]

It's hard to disentangle cause and effect here. Obesity contributes to poverty, and then poverty makes it harder to escape obesity. We know that across the world, lower socioeconomic status is associated with higher rates of obesity. Alexandra Brewis of Arizona State University has identified a layering of stigmas among the poor and obese in America: a layering of "stress, suffering, lost opportunities, downward mobility" that traps people in lives and in bodies that they never chose.[32]

If the aim of stigmatizing people with obesity is to transform them into thin people, it's spectacularly counterproductive. The trouble is that most of us have been so conditioned by weight stigma that we can't even recognize there is anything wrong with it.

Think back to the last newspaper story you read about the obesity crisis. Can you picture the photo that accompanied the story? I will lay good odds that it depicted a person with obesity in a prejudiced and unflattering light. You probably couldn't see the person's head, only the lower body, which was bulging out of a too-small chair or a too-tight pair of jeans. He or she was likely eating something like a supersized hamburger dripping with sauces. Whether you are obese or not, such a photo is designed to trigger feelings of disgust about excess weight. Analysis by the Rudd Center in the United States found that 72 percent of photographs accompanying online news stories about obesity were stigmatizing. The Rudd Center has created its own gallery of images and video clips showing people with obesity in a nonbiased way and in a variety of settings: in the workplace, for example, or grocery shopping for fresh produce.[33]

But more respectful depictions of people with obesity are unlikely to become the norm until there is more collective recognition of how harmful weight stigma really is. There is still a very widespread view that the best way to counteract the rise of obesity is to shame the fat into being less so. These beliefs exist even among policy makers and healthcare professionals.

In 2008, a Mississippi State House bill was proposed to prohibit restaurants from serving food to any person who was obese. In the end, the bill didn't pass, but the very fact that it was mooted shows the extent to which weight stigma is still—despite all the evidence to the contrary—regarded as a useful public health tool.

Among the many other harms caused by fatphobia is the fact that it disincentivizes people from modest weight loss. If you are a person with obesity, many health benefits can be seen from amounts of weight loss as small as around 10 percent of total body weight. This level of weight loss is associated with improved outcomes for type 2 diabetes, hypertension, and heart disease. But viewed through the lens of weight stigma, such modest diet changes may be perceived as worthless, because the individual in question—though healthier—will still look too large to conform to society's idea of an acceptable body.

The persistence of weight stigma may be yet another sign of a society eating in a state of dislocation. The way that we eat has changed too fast for our moral value system to adjust. Without radical changes to our food system, there will be no reduction in the prevalence of obesity. It ought to be a matter of basic human decency, as much as anything, that our culture doesn't treat anyone as subnormal because of their weight. Werner Cahnman addressed this question back in 1968. The answer to how to engage with obesity, Cahnman said, was "an agreement of mutual respect for the common humanity of each and every one of us." Weight stigma, he pointed out, cannot be removed except by treating individuals with obesity as normal human beings—as intelligent and capable as anyone else—and removing any sense of moral shame about their condition.[34]

Fifty years on, we still have not learned this lesson.

CHAPTER THREE

EDIBLE ECONOMICS

IN THIS WORLD, THERE IS NO SHORTAGE OF EXPERTS telling us that we, too, could lead a long and happy life—and maybe even stave off death—if only we made better food choices. Diet books urge us to make "smart" food swaps (although the idea of "smart" in these books can be bizarre to say the least). Nutritionists tell us to make healthy dietary choices such as eating two portions of oily fish a week. The food industry, likewise, promotes the notion that what we eat is purely down to individual choice. If you happen to buy a 10.5-ounce family-size bag of chocolates and scoff the lot rather than dividing it up into the recommended twelve portions, that is your free choice.

Trying to make better food choices is an entirely worthwhile endeavor, whether we define eating better in terms of individual health and pleasure or in terms of supporting more sustainable agriculture. From childhood, the one great power an eater has is deciding whether to open or close his or her mouth to a given food. Thousands of deaths from chronic disease every year could be prevented if everyone managed to choose a healthier way of eating with fewer refined grains, sugary drinks, and portions of processed meat. When enough people demand different ingredients in different quantities, then the whole food system is forced to adjust to accommodate our wishes. It is often said—rightly—that every time we buy food at a farmer's market or through a local independent shop or organic vegetable box scheme instead of a supermarket, we are voting with our forks for an improved food system.

But it's worth remembering that what we eat is never just a matter of personal desire or demand. Or to put it another way, even our desires are shaped by the world around us—by the quantity of foods that are supplied to us, by their cost, and by the stories we are fed about them, often through advertising. Over time, we learn to want this or that food; however, this learning is not mostly determined by the needs of our bodies but by the limits and possibilities of the food supply.

One recent book on food policy argued that as citizens, every time we select a food, we should not only weigh the price, quality, and convenience of a product but also consider larger values such as health and sustainability. Whoever wrote these words clearly never wheeled a trolley through a crowded supermarket on a Saturday morning, fretting about whether to put that week's groceries on the debit or credit card.[1]

Behind the choices an individual makes about food, there are a series of economic circumstances that none of us asked for. It's hard to vote with your fork when you are trapped working in a call center with no fresh food shops nearby and nothing available to buy for your lunch but a sandwich or snacks from a vending machine. Much of what we consume is virtually pushed down our throats by forces of supply over which we have no control and of which we are only dimly aware. On all sides, our food choices are shaped and constrained by economics.

Average incomes are higher and food prices are generally lower than for previous generations, yet this new prosperity has not translated into higher-quality diets. As we'll see, the relative price of goods over the past two decades has pushed consumers strongly in the direction of ultra-processed foods, meat, and sugar. Equally, the economics of modern food has made it less likely that most people will buy good-quality bread and green vegetables. Our diets are shaped by economic policies that have encouraged the massive oversupply of certain ingredients that we consume, almost despite ourselves, without recognizing what we are doing. A case in point is refined vegetable oil, whose ubiquity in modern food is mainly a story of supply-side economics.

A HIDDEN SEA OF OIL

Some of the global changes in diet in recent years are easy to spot, such as the spread of Cavendish bananas across the world. But many of the biggest changes in our diets are hidden from us. If you'd asked me what food has increased by the biggest proportion in the Global Standard Diet since the 1960s, I would have guessed without hesitation that it must be sugar. But it isn't. There is another change to the world's diet over the past fifty years that is even bigger than sugar and yet it has gone almost unremarked on in daily conversation. It is the rise of refined vegetable oil.

General opinion says that what's wrong with modern diets is that they are low in fat and high in sugar. This is true enough, except for the low in fat part. We are certainly drinking far less whole milk and far more skimmed milk than previous generations did. We also eat far smaller amounts of saturated fats such as lard and ghee, and we tend to buy many more low-fat products from the supermarket (many of which are highly processed and laced with sugar). But that does not mean that the average person is following anything like a low-fat diet. To the contrary, one of the early signs of the nutrition transition wherever in the world it has happened is a massive rise in the supply—and hence the consumption—of cheap vegetable oils.[2]

Many consumers have stopped worrying about fat and started worrying about sugar instead, including pseudo-sugars such as high-fructose corn syrup. High sugar consumption is—no question—a problem. The average American diet contains more than three times the World Health Organization's recommended daily dosage of sugar (and this is before we have even factored in hidden natural sugars in such products as fruit smoothies). There is mounting evidence that high-sugar diets are heavily implicated in the epidemic of type 2 diabetes that is sweeping the globe, affecting people in poor countries as well as rich ones.[3]

But if you look at which crops have actually increased most in global abundance over the period that diet-related ill health has become such an epidemic, sugar doesn't even come close. As of 2009, the Global Standard Eater had access to 281 calories a day from sugar and sweeteners, whereas in 1962 the average was 220 calories. This sounds like a big increase until you compare it to the growth in availability of oils over the same period. In absolute terms, the availability of sunflower oil has gone up 275 percent in fifty years while that of soybean oil has increased by a staggering 320 percent over the same period. This figure does not even include the millions of pounds of soybean oil that go to make up animal feed. Admittedly, not every drop of oil will be consumed. When oil is used for deep-frying, much is discarded. It's been estimated that 4.2 liters of waste cooking oil are generated each year per person in the UK, some of which goes to biofuels. But even allowing for wastage, the rise in use of oils has been staggering. As a point of comparison, it's worth noting that the availability of sugar and sweeteners has grown by just 20 percent over the same fifty-year period. Pretty much everywhere, by far the biggest increase in calories in recent decades has come not from sugars (of whatever kind) but from soybeans (which are primarily eaten as soybean oil), followed by palm oil and sunflower oil: all cheap, refined vegetable oils.[4]

Refined vegetable oils, with soybean oil at the top, have added more calories to the world's diet than any other food group, by a wide margin. I've met many people who aspire to eat sugar-free diets and others who worry about trans fats, but I have seldom met anyone who said that their resolution was to cut down on their intake of soybean oil (or other refined vegetable oils) per se. Certainly, there are those who avoid palm oil because of the environmental impact of rainforests being razed to make way for a monoculture. But the vast role that oils play in our diets goes largely unseen and uncontested.[5]

We speak of hidden sugars (such as the unexpected glucose syrup in a pizza topping or the surprising amount of sugar in teriyaki sauce), but oils are still more deeply hidden in our diets. Most of us are well aware that we eat a lot of sugar precisely because we love it so much. We can see the sweetness gleaming at us in the shiny slice of chocolate gâteau, the scoop of praline ice cream, and the crunchy handful of M&Ms. No one deliberately seeks out foods because they are oily, but often we consume them without realizing the oil is there.

In the 1980s, when my father wished to express disapproval of a dish, the word he reached for most frequently was *oily*. When our family bought a takeaway curry from our local Indian restaurant, he would sometimes comment that the chicken biryani or the poppadoms were "not too oily." This was high praise. Oiliness to my father signified something made without care by a cook who could not be bothered to skim away the excess fat on the surface of a dish of Irish stew or to blot the underside of a fried egg before serving it.

I think of my father now because the oiliness he so feared is everywhere. Oils are what plump up muffins and give crispness to fried chicken. But these modern oils are stealthy ingredients that do not advertise their presence. Colin Khoury of the FAO wasn't expecting oils to be such a big part of the Global Standard Diet. When he and his colleagues started to analyze the world's diet through crops, he—like me—expected that the big story would be sugar. Instead, Khoury found that the greatest movement in the data had been toward a handful of refined vegetable oils, with soybean oil at the top of that list. The rise of soybean oil "was a big surprise for me—for sure!" he comments.

Soybean oil—along with other refined vegetable oils—has become an ingredient that sneaks into our mouths almost without us realizing, through foods both healthy and unhealthy. Soybean oil is in some ways the ultimate modern food for Global Standard Eaters. It's cheap, it's abundant, and it lends itself to the creation of thousands of other industrial foods. It has become the seventh-most-eaten food in the world without anyone ever really desiring it.

You would not think that soybean oil would have ever become a food with mass appeal, for the simple reason that it tastes pretty bad. An industry guide to oils and fats from 1951 noted that soybean oil had a tendency to "flavour reversion." What this bland phrase means is that if you don't get it super fresh, this oil will pick up grassy or hay-like odors (if hydrogenated) or taste chemical and fishy if it's unhydrogenated. Soybean oil is no match in frying flavor for the animal fats such as tallow or lard that were extensively used until worries about saturated fat took hold in the 1980s. Slightly fishy-tasting vegetable oil sounds like the last thing anyone would want to eat.[6]

But when it comes to global food markets, price often trumps taste or nutrition. The widespread use of soybean oil started in the United States in the 1940s precisely because its variable flavor quality made it significantly cheaper than its main rivals, peanut and cottonseed oil. Its cheapness created an incentive for manufacturers to opt for soy in food products such as crackers and confectionery, as well as multiple varieties of fried foods, even when other, superior oils were available.

The fact that we eat so much soybean oil shows the power of global supply chains to alter what we eat. Corinna Hawkes is a British food policy expert who first became interested in soya in the early 2000s when she noticed, as Colin Khoury later did, that refined oils had added to our average calories more than any other food. Hawkes had heard food policy colleagues arguing that the rise of soybean oil was simply a function of globalized fast food. There was an assumption that we ate more oil because we demanded more oil, in the form of chips and other fried foods. "But I don't like it when people make assumptions about food," Hawkes coolly remarks. She wanted the evidence.

I meet Hawkes for a lunch of baked eggplant in tomato sauce in a neighborhood Italian café not far from where her daughter goes to school. Hawkes tells me about a year she spent working and researching in Brazil, which brought home to her the unfair fact that millions of people in the world do not have the same easy access to healthy foods that the middle classes have in wealthy Western countries. Hawkes genuinely prefers vegetables to meat and has no great fondness for junk food. "But to eat the way I do in Britain was so expensive when I was in Brazil," she remarks. She takes another forkful of eggplant in its tasty sauce of tomatoes, garlic, and olive oil. In Brazil, she found that middle-income consumers were driven to buy heavily processed foods because they were affordable. Even when making home-cooked meals, the balance was skewed toward far more refined cooking oil than people would have used in the past.

Hawkes was startled to see how much oil home cooks in Brazil would use when making the national dish of beans and rice. "People just pour it on because it's so cheap. All those added calories!" Meanwhile, for her research into the world's diet, she kept finding startling statistics on the growth of oil crops since the 1980s. "I could see that soy oil had gone up by a lot and became very interested in looking at the specifics," she tells me. "I asked, 'Who is producing this oil?'"

Delving deeper, Hawkes found that the rise of soybean oil could be clearly traced to a series of economic policy changes in Brazil. Soybeans are far from a traditional food in Brazil. By custom, it is a land of beans, corn, cassava, and rice, whose export foods were sugar, coffee, and chocolate. In the 1950s, Brazil could not even produce enough soybean oil for its own domestic consumption. The most popular cooking fat at that time was lard. But this all started to change in the 1960s, when new food policies dramatically pushed soybean cultivation. At first, the rise of soy was a kind of accident. Brazil was suffering a food supply crisis with millions going hungry and acted to subsidize the wheat industry. A side effect of all this wheat was to increase the amount of soybeans planted, because Brazilian farmers used soya as a complementary crop to improve the soil for the wheat in between plantings. No one predicted how successful Brazilian soy would become in its own right.[7]

By the 1980s and 1990s, the Brazilian government realized that soy could boost the national economy as a valuable export and introduced a raft of measures to promote it. The government lowered import taxes on fertilizer (soy is a very fertilizer-hungry crop) and removed restrictions on foreign investments in domestic agriculture. A soybean export tax was eliminated. The upshot of all this was that the world food markets were suddenly flooded with cheap soybean oil. It helped that the high season for soya in Brazil coincided with the low season for soya in the United States. From 1990 to 2001, soybean oil production in Brazil increased by two-thirds and exports doubled.

Soybean oil is typical of the modern food economy in that it is mostly grown in one place and eaten in another. Now, most Brazilian soybean oil is used by consumers far away in China and India. As middle-class incomes and populations started to grow in Asia, one of the first things that people spent their money on was more cooking oil. Between 1989 and 1991, China imported nearly two million metric tons of soybean oil. Ten years later, it was closer to fifteen million.[8]

Where are all these spoonfuls of soybean oil going? Hawkes points out that much of it dribbles almost unseen into cheap takeaways made by street vendors. But it is also used by restaurant cooks and home cooks. "I mean, have you *seen* how much oil cooks use in China?" I reply that I haven't yet been to China. "Well, it's a lot," she says, with considerable emphasis.

A year later, when I traveled to Nanjing, I thought of my conversation with Hawkes at every meal. I was struck that even a deliciously simple stir-fry of eggplant and green beans ordered in a restaurant would leave behind a shiny slick of oil on the plate. As the price of cooking oil came down and incomes rose, suddenly it was something the cooks of the world would slosh into the pan more generously than

Figure 3.1.   Cooking oil prices, China, 1991–2006.

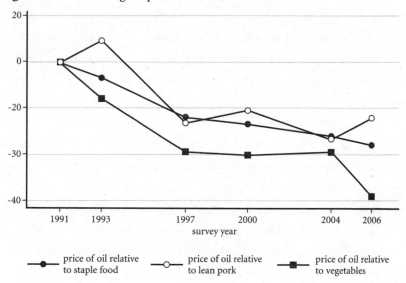

Source: Figure 1 in Lu and Goldman, 2010. Steve Wiggins and Sharada Keats, 2015, "The Rising Cost of a Healthy Diet," Overseas Development Institute, London.

before. For hundreds of years, cooking oil in China—as in most other countries—had been a luxury, something that was used sparingly, by the drop. Rapidly and almost imperceptibly, eaters became accustomed to dishes with an oilier, richer taste than in the past.

Our food may not taste oily to us, but much of it is. One reason we don't recognize oil's presence in our diets is that so much of it goes not into delicious cooking—as in China—but into the food-processing industry, where it is transformed into heavily marketed cereals, biscuits, ice creams, and snacks, allowing the creation of foods that we had no reference point for judging the oiliness of, because they were nothing like the food of the past.

NEVER TASTED BEFORE

Consider a pack of instant noodles. Wherever you live, you can pick one up easily from any supermarket, or if you want a bigger range, you could head to a Korean or Chinese food shop. In every continent of the world, instant ramen or pot noodles—a block of precooked noodles to which you just add hot water—is the way many people manage to eat a hot lunch on a low budget. You can choose them in a hundred different flavors to suit your whim: chow mein, chicken and mushroom, black garlic, pork ramen, extra spicy Thai-style tom yam.

Like so many modern food products, instant noodles—which are beloved, in a faintly ironic way, by many Western food writers—look as if they are offering us a whole rainbow of options. Yet when you look at the list of ingredients, they are basically always the same: wheat, salt, and various flavor enhancers plus vegetable oil to give the noodles a sheen of richness. This may not matter much if you only eat instant noodles once in a blue moon to take the edge off a hangover, but it's different for the people who depend on them for daily sustenance. Chinese consumers bought 38.5 billion portions of instant noodles in 2016. This looks modest if you compare it with the 46 billion portions they bought in 2013, but it still represented thirty lots of instant noodles for the average person: thirty meals of wheat, soybean oil, and MSG. In nutritional terms, instant noodles are narrowing people's options rather than opening them up.[9]

Instant noodles are now eaten even in places where people have no access to the boiling water needed to make them soft. When culinary researcher Faith d'Aluiso traveled to Asmat in New Guinea for her 2005 book *Hungry Planet*, she met a father and two sons. All three looked seriously malnourished:

> As we were talking the older boy pulled a dry brick of instant ramen noodles out of its wrapper and munched it down. His naked, pot-bellied little brother tipped the ramen's flavouring packet into his own mouth and worked the powder around with his tongue until it dissolved. I was mesmerised. I saw this scene play out again and again during our time in Sawa, a place with next to no connection with the rest of the world—children

eating an uncooked convenience food intended to simplify the busy lives of people very far away.[10]

These changes did not happen by accident. Between the 1980s and the present day, there has been a dramatic increase in the amount of foreign direct investment in the food of developing countries. As sales of snack foods approached saturation point in Europe, Canada, and the United States, the multinational food companies looked to opportunities in new untapped markets. Meanwhile, governments in the developing world were desperately looking for new investors to help their economies grow. It seemed like a perfect match.

Foreign direct investment (FDI), a form of investment across borders, happens when a business in one country acquires an interest in the business of another country. From 1990 to 2000, the amount of total FDI into developing countries grew sixfold, from $200 billion to $1.4 trillion. This was a vast influx of much-needed capital into some of the world's poor and middle-income countries. FDI also created thousands of new jobs. By 2004, Walmex—the Mexican version of US supermarket chain Walmart—was the country's biggest private employer, with 109,075 employees.[11]

But the downside of FDI for food was that most of this new money from overseas was going into companies creating products that contributed to obesity among poor and vulnerable consumers. When it came to food, the vast majority of the FDI money went into companies that made ultra-processed items such as breakfast cereals, snack bars, sugary drinks, and chips. From 1980 to 2000, FDI from the United States into foreign food processing rose from $9 billion to $36 billion. Most of the investment came from transnational food companies such as Nestlé and PepsiCo who set up their own factories in countries such as Mexico and Colombia.[12]

Long before FDI came on the scene, many developing countries had their own versions of packaged foods: their own sweets, their own tooth-rotting drinks, their own salty snacks, their own refined vegetable oils and bags of sugar. The difference was that the national companies making these products did not have enough capital to expand very fast. With new money from FDI, the facilities for both manufacturing and marketing these foods rapidly expanded. When moving into Latin America, some soft drink companies explicitly set the target that any given person should never be more than one hundred meters away from the nearest outlet selling one of their drinks.

Thanks to FDI, markets were flooded with foods such as instant noodles that most people in these countries had never tasted before. Investors—quite naturally—are always looking for a return on their investment, and the biggest profits were to be made in processed foods, where the cost of ingredients may be a tiny fraction of the retail price. The FDI system has hugely increased the availability of processed foods for millions of people.

Take Mexico. It was only in the 1980s that market integration between Mexico and the United States began, a process accelerated by the North American Free Trade Agreement in 1992. Three-quarters of FDI food money in Mexico went into processed foods. As rates of obesity started to rise along with the availability of these new foods, there were opportunities for further investment. The Mexican market was now flooded with a new panoply of diet products. Coca-Cola introduced no fewer than twenty new artificially sweetened health drinks into the country in 2005. Whether these really deserve the word *health* on the label is debatable, given increasing evidence that artificially sweetened drinks are, like sugary ones, associated with type 2 diabetes.[13]

We speak of having better food choices, yet for the most part, we eat the foods that food companies want to sell us. People in Mexico did not wake up one morning and decide en masse that they would eat the same foods. Yet that is what has happened, to a remarkable extent, and one of the reasons is that Mexicans—like people almost everywhere—eat so many foods that have been processed out of all recognition from their agricultural state.

FDI is one element in a larger global economy of food that is pushing us to eat fewer whole fresh ingredients and more foods that are heavily processed. It was Carlos Monteiro, a professor of nutrition in Brazil, who suggested that it now makes sense to talk of food as belonging to one of four distinct groups, depending on the degree to which it is processed. In the early 2000s, Monteiro invented the NOVA classification, a new way to think about different foods that gets beyond the old arguments about macronutrients such as fats and carbohydrates. Group 1 in NOVA is made up of whole foods such as fruits, vegetables, and nuts; fresh meat; and plain yogurt. A pedant will point out that even these foods have been processed to some degree before they reach the shops. A carton of milk will have been pasteurized and refrigerated; nuts are shelled. Group 1 also includes foods that are dried or frozen, such as dried mushrooms or frozen peas. But the original ingredient is still quite recognizable: a carrot, an olive, a lamb chop, a potato, a sprig of thyme, a handful of dried cannellini beans. You can see it and name it.[14]

Group 2 foods are what Monteiro calls "processed culinary ingredients." These include butter, salt, oils, sugar, maple syrup, and vinegar. Traditionally, the role of these ingredients was to be used in relatively small quantities to prepare, cook, and enhance foods from group 1.

The third group in Monteiro's classification are simple processed foods. Most of these are made by adding a group 2 food to a group 1 food and heating, fermenting, or treating it in some way. An example would be cheese, made by adding salt and rennet to milk, or pickles, made from adding salt and vinegar to vegetables. Other examples of group 3 foods would be canned fish, beans, or tomatoes. Personally, I'm a big fan of this kind of processed food. With a can of tomatoes, a packet of pasta, and a hunk of parmesan, I can get dinner on the table even when it looks as if there is no food in the house.

In our times, however, a fourth group of foods has started to dominate our plates. These items, taken together, represent a radical reinvention of food, and not for the better. The main use for group 2 foods is now, as Monteiro says, to provide the raw material for yet another group of foods, the ultra-processed ones. Ultra-processed foods, in Monteiro's definition, are "basically confections of group 2 ingredients, typically combined with sophisticated use of additives, to make them edible, palatable and habit-forming." Although they may be marketed as "all-natural" or healthy, they have little in common with group 1 foods. Group 4 foods are low in nutrients and fiber and high in sugars and fats and tend to be based around a very narrow range of basic components, dolled up with colorings and flavorings. Among countless others, group 4 foods include ready-to-heat meals and slimmer's meals, carbonated sweetened drinks and cereal bars, chicken nuggets and hot dogs, sweetened breakfast cereals and "fruit" yogurts, and almost any kind of bread you can buy in the supermarket. And instant noodles.

It is Monteiro's contention that ultra-processed foods of all kinds, more than any single nutrient, are responsible for much of our diet-related ill health. Not everyone in nutrition circles agrees with Monteiro, by any means, but there is a gathering body of evidence supporting his thesis. In 2018, a large population-based study from France found that a 10 percent increase in consumption of ultra-processed foods was associated with a 10 percent higher overall cancer risk, a pattern of eating that was also linked with a higher risk of breast cancer.[15] The study looked at everything from mass-produced breads and chicken nuggets to packet soups and shelf-stable ready meals. Participants in the study—more than one hundred thousand of them—were asked to keep multiple sets of twenty-four-hour food records, which were then verified against blood and urinary biomarkers. These food diaries were repeated every six months for eight years and compared with self-reported cases of cancer and medical records. As nutrition studies go, this was a very strong and sizeable one.

What isn't yet clear is exactly where the risk of ill health as a result of consuming large amounts of ultra-processed foods comes from (or where the cut-off point for safe levels of consumption might be). Some group 4 foods contain additives that are known to be harmful to human health in large doses, such as the sodium nitrite and nitrate in processed meat such as bacon.[16] These interact with certain compounds in meat to form N-nitroso compounds, which cause cancer. But there are no N-nitroso compounds in ultra-processed cakes and desserts, among others. Another theory is that some of the processes involved in making and selling these foods—such as the application of very high heats and the use of plastic packaging—can generate carcinogens. A more obvious explanation is that group 4 foods tend to have a nutritional composition that is very low in fiber and vitamin density and high in fat, sugar, and salt. The most popular ultra-processed foods consumed in the French cancer study were sugary foods (26 percent of all ultra-processed foods consumed) and sugary drinks, followed by breakfast cereals, which yet again are sugary. Ultra-processed foods contain far larger quantities of ingredients such as refined oil and sugar than you would probably add in your kitchen at home (although as we've seen, our ideas of what is a normal amount of oil to add to a dish have also inflated).[17]

A significant problem with group 4 foods from the point of view of consumer health is that they cannot be altered. With a jar of sugar or a bottle of oil in your hand, you as the cook are free to use as much or as little as you like. You might decide to use less sugar in a cake recipe or a dash less oil in a stir-fry. When you are presented with sugar and oil in the form of a doughnut or a frozen pizza, the decisions about how much of each ingredient to add have already been made for you. Your only choice is which brand to buy and how much of it to eat.

Given the choice, it is probably not the best move to consume too many ultra-processed foods. But it is an extremely smart move to sell them, which explains why these foods are so widely available in our shops. In Canada, ultra-processed foods already made up around a quarter of the average food basket in 1940; today, they form more

Figure 3.2.   Rising consumption of processed potato products, 1945–1970.

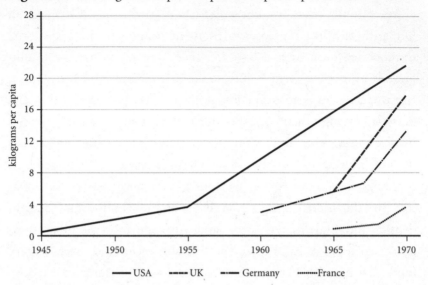

Source: Organisation for Economic Cooperation and Development, *The Impact of Multi-national Enterprises on National Scientific and Technical Capacities: Food Industry* (Paris OECD, 1979), 118.

than half. Behind the global rise of ultra-processed foods, there is a stark piece of economic arithmetic. Group 4 foods create much bigger profit margins than those in group 1. Whole foods tend to generate profits of around 3–6 percent, whereas ultra-processed foods—made from cheap ingredients and produced on an epic scale—can yield profits of around 15 percent.[18]

We eaters are bearing the cost of a food economy in which ultra-processed foods are so cheaply and readily available. Carlos Monteiro argues that ultra-processed foods are best avoided altogether. To me, this is a counsel of perfection. I don't believe that you will terminally damage your health because you occasionally eat a bowl of honey-nut cornflakes or—God forbid!—take a sip of Coca-Cola.[19]

But when ultra-processed foods start to form the bulk of what some people eat on any given day, we are in new and disturbing territory for human nutrition. More than half of the calorie intake in the

United States—57.9 percent—now consists of ultra-processed food, and the United Kingdom is not far behind, with a diet that is around 50.7 percent ultra-processed. In a market stuffed with ultra-processed foods, many of our food choices consist of fairly meaningless decisions between this or that version of what is essentially the same product. If Monteiro is right, the real choice is not between the merits of two different breakfast cereals but between a diet skewed toward group 4 and a diet founded more on the whole foods of group 1 (supplemented with the useful processed foods of group 3). As a normal consumer on an average budget, it's hard to make this leap. We have reached the stage where the ultra-processed versions of many foods start to seem like the normal ones. This has even become true for such basic human foods as bread. When bread is made as it always used to be— from little but flour, salt, and yeast—it counts as a group 3 processed food, according to Monteiro. But when bread is made using sped-up industrial methods with emulsifiers and other additives, bread itself becomes a group 4 ultra-processed food. And good luck with finding any bread in the average supermarket that *isn't* ultra-processed.[20]

THE PRICE OF BREAD

Why do the rich in so many countries eat such bad bread? The poor mostly eat bad bread too, but they have little choice in the matter. The real puzzle is why the wealthy—who could easily afford to buy a wholesome loaf of substance that actually tastes of something— should often be content with bread of mediocre quality.

It is a curious fact of modern affluent societies that we start to eat less bread as we become richer and, stranger still, pay less attention to the quality of the bread that we do eat. It's one thing that factory-made "bread" should be high in salt; the really odd thing is that it's also high in sugar, not to mention padded with "dough conditioners" and preservatives and given hardly any time to prove before it goes in the oven, regardless of whether it is called farmhouse or lower carb or wholemeal. And yet we accept this substance as bread. Maybe we even start to prefer it, because it's what we know.

Like much of what we eat, sliced industrial bread is a compromise. But at any given time, we compromise more on certain foods than on others. And the choices that we make with our money tell us something about what our culture holds to be important. Take fresh blueberries. These used to be a luxury item, whereas now sales of fresh berries—despite their expense—have overtaken sales of apples and bananas in the United Kingdom, pushed along by the vogue for making smoothies in high-speed blenders and by the blueberry's status as a superfood. Berries have gone from luxury to staple for many people. Bread, by contrast, is in danger of losing its centuries-old status as a staple. Ours is a culture where many people obsess more about the quality of a post-gym snack than they do about the quality of bread, that most basic form of sustenance. The devaluation of bread speaks of a culture that has stopped seeing food as a fundamental need and started seeing it as a kind of leisure activity.[21]

The nutrition transition, as we have seen, is largely driven by prosperity. But that prosperity, strangely, does not always translate into a better quality of staple foods but the opposite. You might think that being richer—as a nation or as an individual—would naturally lead to the consumption of better-quality and more nourishing foods. In most countries, however, a rising standard of living has gone hand in hand with diets rich in quantity but poor in quality. Like our great-grandmothers, we value celebration foods such as meat and fruit and sugar more highly than boring staples such as bread and rice. Unlike our great-grandmothers—who cherished every grain of rice—we can afford so many feasting foods that we start to neglect the basics.

Some might say that we're right to prioritize other foods over bread, given that protein is a more essential nutrient than carbohydrate. But it is not as if we are currently eating anything like a low-carb diet. Even as bread is overlooked, the amount of grains in the diet of the

Global Average Eater has actually gone up—from 976 calories of grains a day in 1961 to 1,118 calories in 2009. There is no shortage of refined wheat in our diet, from breakfast cereals to pasta.

Not everyone in rich societies, needless to say, eats ultra-processed bread. Across the world, a new movement of keen hobby cooks are starting to teach themselves the skills of baking sourdough loaves from scratch, and there has been a modest revival of artisanal bakeries selling slow-risen bread, made from nothing but flour, water, salt, yeast, heat, and a baker's touch. At just the point that mainstream bread became more or less debased, sourdough toast was reinvented as something that some people will spend a fortune on, especially if it is topped with small-batch almond butter or some variation of smashed avocado. But such bread is far from the norm, and "hipster sourdough" has become an object of mockery—as if there were something pretentious about paying too much attention to the quality of bread.

Minding about the quality of bread used to be something as universal as breathing. In 1850, every person in London ate the same bread, and everyone in the city was a sourdough snob. As George Dodd noted in his panoramic book *The Food of London* in 1853, "A Loaf of wheaten bread is a London staple; [a worker] demands it as well as a peer." Everyone knew that the standard loaf of London was a "quartern loaf" weighing around four and a half pounds, much bigger than a modern loaf. Despite London's modernity, bread was still made by hand, kneaded by a man "straddling and wriggling on the end of a lever or pole." Germans in London, such as the chemist Frederick Accum, found London bread to be of poor quality—routinely adulterated with alum, a chemical to improve the rise. But compared to modern British supermarket bread, it was still a hearty, crusty loaf, made from a slow-acting ferment of boiled potatoes mashed with yeast. There were 2,500 bakers making and selling quartern bread in London in 1853. It was the most basic way for a Londoner to satisfy hunger, a single flavor and texture uniting rich and poor.

If someone asked us to give them London bread today, where would we direct them? Would we offer them the chewy-warm bagels of Brick Lane or the flat moreish chapattis of Drummond Street, near Euston? If we took them to Borough Market, we could feed them loaves lovelier and purer than those Victorian quarterns. There are now artisanal bakers in London who make sourdoughs of better quality than London bread at any time in the past: ovals of rye and long whole-wheat sticks studded with currants and hazelnuts, domes of sturdy white sourdough with concentric floured circles on top. But at five dollars and upward for a loaf, we couldn't honestly pretend that such bread is a true London staple. For most of the eight million who live and work in London, rich as well as poor, bread now means an industrial sliced loaf bought from the local supermarket, flabby, hardly proved, and packed with additives: the same unsatisfactory bread found in supermarkets everywhere.

Economists refer to bread as an inferior good, meaning that it is valued and desired less as people become richer. The same goes for potatoes. Inferior goods are ones whose value falls as income rises. Demand for these starchy staple foods invariably declines as people get richer. Bread consumption has plummeted in all the rich countries of the world. In the United Kingdom, bread consumption effectively halved from 1880 to 1975. The less we ate of it, the less we seemed to care about it.[22]

Bread used to be central to our lives, largely because it gobbled up so much of our incomes. In the nineteenth century, an average British farmworker in the county of Somerset spent more than twice as much on bread as on rent (£11 14s per annum as against £5 4s). With industrially produced bread and rising incomes, it is now more than possible for a family to spend less each month on bread and butter than on mobile phones and Wi-Fi. Bread's lowly status can be seen from the fact that so much of it is discarded. Our ancestors used up every scrap of a loaf and put any stale bits to good use: think of the thick, oily bread soups of Italy and Portugal or the bread stuffings of America. Nowadays, old bread, sad to say, seems almost worthless (partly because those industrial loaves do not age gracefully, going straight from soft and fresh to soft and moldy). In the United Kingdom, bread is the single most wasted food, with 32 percent of all loaves purchased thrown away.[23]

The bread culture is changing even in places where good bread used to be the bedrock of an entire cuisine. Take rye. This dark, nutty grain was once a beloved daily staple in the Czech Republic (or Czechoslovakia as it used to be). Rye has been cultivated in central Europe since the Middle Ages. Having a taste for rye was part of what it meant to be Czech, just as part of being Italian is enjoying durum wheat pasta. The standard Czech bread was a rounded sourdough made from half wheat, half rye, flavored with caraway seeds. Hefty slices of it were eaten at every meal—and when the loaf started to go stale, the last dark chewy crumbs were used up in soups, with mushrooms and dill.

Since the 1960s, rye consumption in the Czech Republic has plummeted, as white wheat bread—which is far less nutritious than the old rye loaves—took over. Rye has gone from being a staple food to something that now contributes fewer calories to the average Czech person's diet than fruit. In 1962, the average Czech got 345 calories a day from rye. By 2009, this had dropped to a mere 66.1 calories. This is a huge change—in flavor and culture as much as nutrition. When rye is no longer the basic way to satisfy hunger, the entire palate of the Czech diet changes.[24]

Yet, like other changes to our diets—such as the rise of soybean oil—the fall of rye bread in the Czech Republic can easily go undetected. At a conference in Oxford in the summer of 2017, I got to chatting with a Czech American food writer and historian named Michael Krondl who divides his time between Prague and New York City. I asked Krondl what he made of the fact that Czech people ate so little rye now compared to in the past. Krondl assured me I was mistaken and people in Prague ate just as much rye bread as ever. Krondl himself was a regular consumer of rye bread, and he was convinced that other Czech eaters continued to share his tastes.

A few days later, Krondl emailed me to say that he was wrong about what he called the "rye question." He had spent the day wandering around Prague bakeries and supermarkets and found to his surprise that wheaten rolls did indeed outnumber rye bread by at least four to one. Another discovery was that even the supposed rye loaves in Prague now often had some cheaper wheat flour mixed in. What's more, Krondl had done some searching for figures on rye from the Czech statistics office that confirmed that a significant shift away from rye had happened after the end of the communist state of Czechoslovakia in 1989. "No mystery there," he wrote. Before 1989, bread was state controlled, whereas from the 1990s onward, people bought their bread from private bakeries. These bakeries preferred to sell white wheaten rolls (*rohliky*)—sometimes sprinkled with sesame and poppy seeds—because wheat flour is far less demanding to work with than rye. Meanwhile, consumers who had been stuck buying rye bread for so many years under communism happily made the switch to this lighter bread, even if the quality was not, in fact, as good.

Much of what we think of as taste is actually economics, as the case of bread illustrates. In many ways, we should count ourselves lucky to care so little about bread. The lowly status of bread in our culture is a sign of how little we now depend on it. Through the centuries, bread consumption has always fallen whenever living standards rose.[25] It would have been a miserable existence to be a British rural laborer of the nineteenth century, spending so much on bread that there was little left in the family purse for anything else, except for tea and treacle (a dark, sticky molasses-like syrup that was cheaper than butter).

What's true of bread is true of every other staple food. We can gauge the prosperity of a nation by what percentage of calories eaten are made up by staple carbohydrates. Based on figures from 2001, the average person in Cambodia got 76.7 percent of his or her daily energy from rice.[26] To depend so heavily on a single crop is to run the risk of malnutrition. Compare and contrast with Spain in 2003, where just 22 percent of the average person's calories were from cereals, with the remainder made up of vegetable oils, mostly olive (20 percent), fruit and vegetables (7 percent), starchy roots (4 percent), animal products (14 percent), dairy (8 percent), sugar (10 percent), alcohol (5 percent), and other foods (10 percent). Given the choice between living and eating in Cambodia or in Spain, who wouldn't opt for the richness and variety of Spanish food? Depending on a single crop such as rice to stay alive is a precarious way to eat as well as a very dull one.

It used to be that you could divide the countries of the world up according to the staple food their people ate. Each country had its own carbohydrate, an inexpensive basic ingredient on which diets and lives centered. As of 2003, when the world population was roughly 6.5 billion, a billion people—mostly in Africa—relied on roots and tubers as their staple food, from cassava and sweet potatoes to yams. A further four billion people relied on rice, maize, wheat, or a combination of the three.[27]

But the remaining 1.5 billion people—including the population of most of Western Europe, the United States, Canada, and Australia—no longer relied on a single staple food. Within these populations, some people ate bread; others ate rice noodles and yet others muesli. But these starchy foods no longer had a unique place in the diet. Now, a decade or more on, millions in Russia, Japan, China, and South America are following suit and setting aside the old staple foods.

Eating without a staple is now becoming the global norm. Moving away from staple foods such as bread or rice is part of the process of distancing ourselves from hunger. It is a great luxury to reach the point where we don't depend on a single belly-filling starch, where whole populations can start to live to eat instead of eat to live.

But trying to eat without a staple food also comes with its own dilemmas. One of the things that we lose when our diet has no staple food is our previous sense of culinary structure, which played out in different ways in different cultures. In France, people used to know that it couldn't be a meal without bread. In South Korea, it couldn't be a meal without rice. What happens—to our psyche as much as our bodies—when food can be absolutely anything? The freedom of choice is extraordinary—and terrifying.

A second dilemma of the diet without staples is that when we can afford not to obsess too much about satisfying our hunger, we cease to value food in the same way and our senses become dulled to changes in its quality. Just as Sherlock Holmes could recognize every variety of cigarette from its ash, an eighteenth-century European could distinguish different types of wheat from a single bite of bread. People knew when a loaf of bread had been made from inferior grain. Now, as chef Dan Barber has observed, we no longer even expect wheat to *have* a taste.

The problem with modern bread is not just how it is made but that the fundamental ingredient—wheat—has become so debased. The typical flour in an American loaf will be pretty old—which degrades the nutrients—as well as bleached and padded with extra wheat gluten to produce a higher yield and prolong the loaf's shelf life.[28]

The downgrading of bread is part of a wider phenomenon that lies at the heart of the food paradox. What's at stake here is the quality not only of bread but of food itself. Because we can now afford to be casual about food, we have forgotten just how valuable it really is. We don't expect to pay much for our meals and we therefore offer the people who work to produce them scant reward. A 2014 report by the New Economics Foundation found that the UK food system employed around 11 percent of the country's labor force but paid them salaries that were under half the UK average. The danger we are facing is that food itself is becoming an inferior good. Spending money is one of the ways that we express what matters to us, and the message we are sending by how we spend our money now is that food doesn't matter very much.[29]

ENGEL'S LAW

Never in history has the average human spent such a low percentage of his or her income on food. As with other aspects of our modern diets, this is both a blessing and a curse. It is an iron law of economics that as income rises, the percentage of it we spend on food drops, even if our food bill is higher in absolute terms. This makes sense. Once people don't have to worry about basic subsistence, there are so many things we might want to buy other than food, from holidays to TVs to smartphones and apps.[30]

The fact that people living in richer countries spend less on food, proportional to other things, is called Engel's law, after the German statistician Ernst Engel (1821–1896). Engel was born in Dresden and spent much of his life studying the way the working classes lived in that city. He noticed that in Dresden, the poorer the family, the more that family would spend on food as a proportion of household income.[31]

Engel found that this rule about food and spending applied not just to individuals but to countries. The more affluent the country, the smaller the share of spending that goes on food. The poorer the country, the larger the share that goes on food.

There are very few laws that all economists can agree on. Humans live in a state of flux, and new situations tend to make a mockery of old economics. But Engel's law has held firm for 150 years. In 2009, it was described by two Australian economists as "arguably the most widely accepted empirical regularity in all economics." It's such a reliable law that it can even be used as a measure to rate the relative poverty of a country. If you know that a country—Madagascar, say—spends 57 percent of its per capita consumption on food, you can be sure that it is very far down the league tables of the wealth of nations.[32]

As of 2005, based on data covering 132 countries, the 16 that spent the highest percentage of their budget on food were all poor countries in Africa, including Guinea-Bissau, Mozambique, Sierra Leone, Togo, and Burkina Faso. These countries were all spending around half of their per capita consumer income on food, rising as high as 62.2 percent for the Democratic Republic of Congo. More recent data from 2015 suggests that many Asian countries also spend a large share of income on food, including Indonesia (33.4 percent), the Philippines (42.8 percent), and Pakistan (47.7 percent). At the other extreme, in the wealthiest countries in the world a minuscule percentage of per capita spending money went on food to eat at home. Japan and Belgium still spent quite a lot on food—14.2 percent—but other rich countries spent a much lower percentage: Australia, 10.2 percent; Canada, 9.3 percent; United Kingdom, 8.4 percent; United States, 6.4 percent. These figures are based not on total income but on consumer spending: everything that a household buys, from cars to clothes and healthcare to electricity. For most people living in this richest group of countries, food now makes up a tiny fraction of the household budget, although these figures do not include food bought away from home, which in the United States accounted for a further 4.3 percent of disposable income in 2014.[33]

There are exceptions to these economic food rules. Not every country spends exactly the amount on food that its national wealth would predict. Within the general tendency of Engel's law, there are a few countries that spend more on food than would be predicted based on average income levels. As of 2005, France had a higher per capita income than Australia but spent significantly more of its consumption income on food (10.6 percent versus 8.5 percent). This reflects the fact that despite the encroachments of the modern world, France still has one of the proudest and most tradition-minded food cultures in the world. It is still—for some, at least—a land of artichokes and food markets, of truffles and raw-milk cheese. Food in France cannot be reduced to numbers on a ledger sheet.[34]

In much of the rest of the world, sadly, it seems that it can. In many affluent countries, we have flipped from spending too much on our food to spending too little. No one—whether a country or a person— would want to be spending more than half of his or her income on food; that would be a sign of abject poverty, of lives dominated by canceling out hunger. The question is whether middle-class households in rich countries now spend too low a percentage of their income on food for the sake of a sustainable food economy or for health.

Figure 3.3.   Average spending on food consumed in the home by country, as a percentage of household expenditure, in 2016 (based on data from the USDA, Euromonitor).

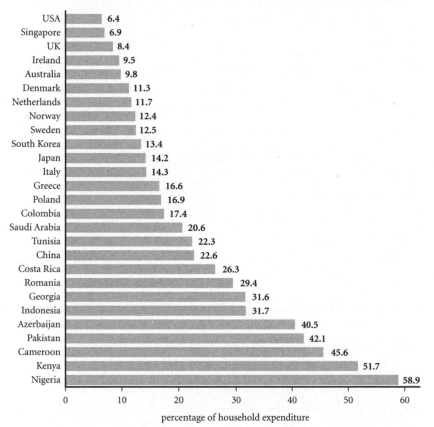

percentage of household expenditure

"I economize on food because it's the one thing in the budget that I can spend less on without impacting the family," remarked an acquaintance of mine who has three children. Many of the other costs of life are fixed and more or less beyond the household's control: the mortgage, gas for the car, school uniforms, insurance, washing machine repairs, five sets of mobile phones, five sets of shoes and winter coats, five sets of everything. But food is one of the few things that can be squeezed here and there to bring costs down. This household buys store-brand teabags, frozen vegetables, and the cheapest chicken instead of free range. When things get tight, the adults may eat peanut butter and jam sandwiches for lunch for a few days using whatever whole wheat sliced bread is on special offer. It's not that they think this bread is good, particularly, but if it will save a few dollars in a stretched budget, it will do. "Food is fuel," she comments.

Spending such a low percentage of our income on food is a kind of progress, but one of the greatest casualties has been food quality. Economists talk about elasticity to describe the extent to which we are prepared to change the quantity of something we buy under changing circumstances, such as when our incomes alter or when prices change. Generally, price elasticity will be negative: when an item becomes more expensive, you buy less of it or switch to buying something else. Food in general is not very price elastic because everyone needs to eat. But certain foods are more price elastic than others. Fruit and vegetables, for example, are very price elastic. In times of inflation, people on low incomes will buy fewer of them, because they don't do much to dent your hunger for the price per pound.[35]

One of the great puzzles of food economics, however, is what happens to our grocery shopping when incomes go up. When a product is of better quality, and we earn more, we should in theory be prepared to pay more for it. Yet with food, this is not necessarily the case. We expect to pay more for a high-definition video on iTunes than a standard-definition one because we can see the benefits in the crisper, clearer image on the screen. So why won't we pay more for a high-quality loaf of bread? Is it because we can't see the benefits, whether in terms of taste or nutrition?

In 2013, an economist studying food found that the quality elasticity of many basic foodstuffs, such as milk, butter, eggs, and bread, was remarkably low. What this means is that most of us in rich countries don't want to pay more for better-quality milk, butter, eggs, and bread if we can manage to pay less for worse-quality ones. The economist's explanation was that items such as bread and eggs "belong to fairly homogenous product groups where it can be difficult to distinguish the quality."[36]

To me, this is a sign of how lost we are when it comes to food. At a sensory level, there's nothing homogenous about the quality of eggs. There is a huge difference in taste, texture, and nutrition between a watery mass-produced egg and one laid by a well-fed free-range chicken, just as there is a huge difference between sliced Wonder Bread and a real loaf of slow-fermented bread made with nothing but flour, salt, water, and yeast. But if most consumers can't tell the difference, why pay the extra, especially when incomes are squeezed and the cost of housing is so high?

When incomes rise, people want to buy the things that prove to themselves that they are living the good life. Sad to say, better bread and higher-quality eggs do not seem to fit the bill. Just because we can afford these foods does not mean that we think they are worth the extra money. The foods that people spend more money on as soon as they have spare cash tend to be the old prestige foods such as meat—the ones that an earlier generation associated with wealth and success, even if they are now cheapened beyond recognition.

According to Engel's law, we spend ever less of our income on food as we get richer. But we also choose *different* foods. Perhaps the most striking addition to our diets with rising prosperity has been meat, especially chicken. Where bread has vanished, white meat has come to take its place. In contrast to foods such as bread, white meat is very income elastic: the richer you are, the more of it you are likely to buy. Eating all this chicken looks like a very modern habit, but our hunger for chicken is something that we share with our great-grandmothers. It's just that we can afford an awful lot more chicken than they ever could.

GIVE US THIS DAY OUR DAILY MEAT

It was after the waiters refilled our plates for the second time that I lost track of the number of dishes I'd eaten. In front of me were at least eight round stainless-steel bowls filled with various stews, dals and other lentil dishes, curries (some mild and some hot), and dumplings soaked in yogurt sauce. There were multiple chutneys, including a fresh green cilantro one and another sharper and chunkier one made from lime. There were some little fried savory morsels shaped like cigars and multiple flatbreads that were replaced with fresh ones as soon as they were eaten. More than that, I can't tell you because after a while, my head started spinning from all the different dishes, though I had drunk no alcohol.

It was the winter of 2016, and I was eating lunch in Mumbai with Indian food blogger Antoine Lewis. We had never met before, but Lewis kindly said that if I waited for him on the corner by the old Regal Cinema near the Gateway of India, he would take me to a place called Shree Thaker Bojanalay for the best vegetarian thali in the city (a thali being multiple small dishes arranged on a circular metal platter to make a complete balanced meal). Our taxi wove down narrow side streets full of people, cows, stationery shops, and fruit sellers before arriving at what looked more like a launderette than a restaurant. Shree Thaker Bojanalay is located on the first floor of an unpromising-looking boardinghouse, but the meal I ate there with Lewis—who blogs as "the curly-haired cook"—was memorable. The flavors and textures were so varied that each bite stayed interesting and satisfying, and I ate long past the point of fullness. The meal happened to be entirely vegetarian, but I swear no meat could have made it any nicer.[37]

I have often thought that being vegetarian would be easy if only I lived in India. Unlike in the United States, where a meatless meal is seen as exceptional, in India, the vegetarian meal is the normal option. In Europe, a vegetarian may be bombarded with intrusive questions about why he or she doesn't eat meat. Is it for the ethics or the taste? How do you get your protein? Aren't you being impolite to refuse a slice of turkey at Christmas? By contrast, in India, it is the meat eaters who are expected to explain their choices, because vegetarian food is the mainstream way of eating. VEG AND NON-VEG was the legend I saw emblazoned on countless restaurants in Mumbai. To be a meat eater in India is to be "non-veg."

In India, many forms of meat are problematic to different religious groups. Most Hindus frown on beef and—under Narendra Modi's nationalist government—the slaughter of cows has been banned in the majority of states. For Muslims, pork is taboo. As for Jains, in addition to avoiding meat, they will not eat so much as a carrot, in case the act of pulling the roots from the ground should upset an insect. Out of these strict religious limitations on meat, Indian cooks have devised a vegetarian cuisine of stupendous diversity. The lentil dishes alone are legion.

Yet despite the glories of Indian vegetarian food—not to mention the many deep-seated religious taboos against nonvegetarian food—Indians are turning to meat as never before. In just a few years, from 2004 to 2010, the average Indian (not that such a person really exists) nearly doubled the amount spent on food, from 1,341 Indian rupees ($18) to 2,508 rupees (around $34). This is a colossal increase in such a short space of time. Market projections suggest that by 2030, Indian cities will be consuming 1,277 percent more poultry than they did in the year 2000.[38]

Nothing in this vast country is simple, but the basic reason for India's newfound meat habit is money. Meat used to be too expensive for all but a small minority, whereas now it has become affordable for millions. There's a direct correlation between bigger incomes and eating more meat. A study showed that for each increase in annual income of $1,000, meat consumption goes up by 2.6 pounds per person in Asian countries.[39]

What India's new meat habit shows is that when it comes to food, economics may trump spirituality. In the past, many observers of India argued that the links between vegetarian food and religion were so fiercely held that the country would never adopt the meat eating of the West. But it turns out that many (though not all) of those who abstained from meat did so only because they could not afford it rather than because of deep-seated beliefs.

Figure 3.4.   The expansion of pork production, 2005–2015.

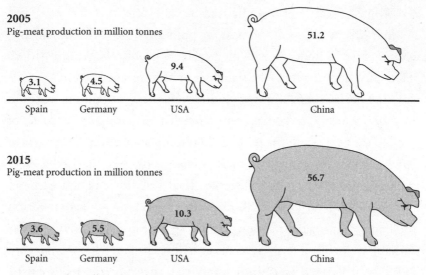

2005
Pig-meat production in million tonnes

Spain	Germany	USA	China
3.1	4.5	9.4	51.2

2015
Pig-meat production in million tonnes

Spain	Germany	USA	China
3.6	5.5	10.3	56.7

Sources: Erik Millstone and Tim Lang, *The Atlas of Food*, Earthscan 2008 and *Statistics 2015 Pigmeat*, Danish Agriculture and Food Council.

There was always rather more meat eating in India than was acknowledged in public, as Charmaine O'Brien, author of *The Penguin Food Guide to India*, has written. O'Brien describes Hindu men escaping to "dark, windowless bars" to share chicken tikka in a kind of macho ritual of rebellion from their wives' vegetarian cooking. The difference now is that for the younger generation of affluent Indians, meat eating has come out in public, and there is just so much more of it than there ever was in the past.[40]

Taking another bite of his vegetarian thali, Antoine Lewis tells me that he has noticed a striking increase in chicken consumption in Mumbai. "In just the past five to ten years, chicken is everywhere." In the evenings, Lewis sees rising numbers of young people going out for fried chicken or Chinese chili chicken, not for the taste, exactly—"the chicken itself doesn't taste of anything, right?"—but just because they can. In a country where family meals are deeply rule-bound, it can feel like liberation to go to KFC or McDonald's.

Someone else who has observed the growing Indian predilection for chicken is economics and food journalist Vikram Doctor, who writes for the *Economic Times*. Over a vegetarian meal of bel poori—deeply flavorful little fried puffs that you fill with spoonfuls of lentils and yogurt—Vikram tells me that he sees chicken as "a non-veg equivalent of paneer" (paneer being a fresh Indian cheese with a blandness similar to tofu). Chicken's success in India is down to the way it can seem like meat that is hardly meat at all, as Vikram discovered when he researched the subject for his Real Food Podcast, broadcast in January 2017, soon after we met. Vikram found that chicken had come to seem so unlike meat that even some people who call themselves vegetarians have started to eat it.

Much as Vikram laments the blandness of Indian chicken, and the poor conditions in which almost all of it is raised, he admits that it may be a useful form of protein in a country where millions still suffer from malnutrition. If you had grown up protein-deprived and poor, why wouldn't you buy a little chicken as soon as you could afford it?

The appeal of chicken stretches far beyond India. If you ask what the world eats for dinner, the answer, often as not, is chicken. "Is it possible, just once, we could get something to eat for dinner around here that's not the goddamned fucking chicken?" asks the bad-tempered grandfather in the 2006 film *Little Miss Sunshine*. Chicken consumption has been growing rapidly almost everywhere. Since the 1970s, global poultry production has more than doubled, and in 2013, chicken became the second-most-consumed meat on the planet, after pork (which remains the preferred meat of the Chinese). Between 2008 and 2013, global poultry exports expanded by more than a quarter as mass-produced chicken started to reach countries in sub-Saharan Africa such as Ghana.[41]

Chicken has become the world's favorite form of protein: inoffensive, supposedly healthy, combinable with almost any flavoring, and available everywhere. Chicken's rise can quite clearly be mapped onto rising economic prosperity. As countries become richer, consumption of poultry meat increases. It increases all the more because as the average person has gotten richer, the average chicken has gotten much cheaper. From the 1970s onward, the global chicken industry became industrialized on a vast scale. In the early 1960s, almost all of the world's chicken was produced in the United States, the Netherlands, and Denmark. But now, Brazil and countries in Asia such as Thailand and China have overtaken the United States as lead chicken producers and are producing it in such volume and with such efficiencies that chicken, once a luxury meat, has become accessible to all except the poorest.[42]

The rise of meat and oil and the fall of bread is one way to summarize the economic story of how we eat now. Our food markets encourage us to treat bread with disdain and to treat the old luxury foods as if they were staples. This story reached an absurdist climax with the launch of KFC's Double Down in 2010. This abomination consisted of a chicken burger, but instead of the chicken being sandwiched in a bun, the chicken itself—two fried breaded fillets—formed the outer casing for the sandwich. Inside the chicken was more animal food in the form of bacon and melted Monterey Jack cheese. The Double Down was a barbecue-flavored, overhyped symbol of how meat has supplanted bread in our lives. Some went so far as to argue that this oily meat sandwich was a healthier option than a normal chicken burger because it was so low in carbohydrate and high in protein, which fits with certain orthodoxies of healthy food.[43]

In many of the impoverished peasant societies of the past, someone was lucky if he or she ate fresh meat more than a few times a year. The slaughtering of a pig was a high feast day. But today, the average Westerner eats more than twice as much meat as bread (which admittedly is partly because other carbohydrates, such as rice and pasta, have spread across the globe).[44]

The shift away from staple foods and toward meat has had great environmental costs that stretch way beyond our dinner tables. A diet based on wheat uses one-sixth of the land required to cater for the kind of meat-rich diets eaten by most people in the United States and

Europe. Largely because of the centrality of meat in the American diet, the United States only uses 34 percent of the food it produces to feed humans directly, because so much of what is grown goes instead—much more inefficiently—to feed animals. It has been calculated that if Spain could return to its traditional Mediterranean diet, abandoning the Westernized additions of recent years, its greenhouse gas emissions would fall by 72 percent and land use by 58 percent.[45]

Are bad bread and cheapened meats the price we pay for living in modern prosperous societies? Or is there another way? Experts in sustainable diets such as Tim Lang, former professor in food policy at City University in London, argue that it is a matter of urgency for governments to promote alternative ways of eating because current levels of meat consumption are unsustainable, particularly considering that by 2050, most estimates say there will be nine billion mouths on earth to feed. Lang is among those who have pointed out that the true cost of cheap meat—and highly processed food in general—is much higher than the price that consumers pay in the shops, once externalities are taken into account. Cheap meat is not cheap if the result is diet-related disease and pollution.[46]

AMONG AFFLUENT URBAN PEOPLE AROUND THE WORLD, there are signs that some eaters are turning their backs on the whole culture of meat and experimenting with new ways of eating, involving a return to staple grains and vegetables.

As populations become richer and escape the horrors of famine, they move away from staple foods in favor of the old prestige foods, including meat. But when people become richer still, and more health conscious, cheap meat tends to lose its former desirability and affluent populations start to enter the fifth stage. Suddenly, some people are prepared to spend more money than you would believe on a tiny bag of mixed seeds, and many of the old staple foods such as millet and spelt have been reinvented as pricy health foods.

Veganism has become remarkably chic of late, a development that can seem surprising to eaters of an older generation. In 2018, at the age of ninety-five, the great African American chef Leah Chase, from New Orleans, expressed her consternation at seeing a sign on a building advertising VEGAN SOUL FOOD. "What the heck is vegan soul food?" she asked. To her, this supposedly fancy new cuisine was no different from the food of poverty that she grew up on. "I was six years old when the Great Depression came, and there was no meat for anybody," she recalled.[47]

It is extraordinary to think how far and how fast the economics of food has changed. Once, to be poor was to be denied meat, except on feast days, whereas now it is often vegetables that are hard to afford. Many consumers are trapped in a situation where the food purchases that make sense in terms of the household budget are not the ones most likely to support health.

VALUE FOR MONEY

When it comes to choosing healthy foods, the dice are heavily loaded against consumers on low incomes. Over the past thirty years, the cost of healthy foods has consistently risen faster than the price of junk foods. Fruits and vegetables have always been expensive to produce: crops such as bell peppers or spinach take a lot of water to grow and are by their very nature costly to ship and store. Food journalist Tamar Haspel has observed that it is unrealistic to expect vegetables to cost as little per serving as grain-based junk foods: "Broccoli just ain't wheat," she notes.[48]

The salient point, however, is not just that vegetables are expensive in absolute terms but that they are much more expensive than they used to be, relative to other foods. In the United States from 1980 to 2011, it became more than *twice* as expensive for Americans to purchase fresh fruit and vegetables compared to purchasing sugary carbonated beverages. Tomatoes and broccoli are far more expensive on average than they

Figure 3.5.   The rising cost of green vegetables relative to ice cream in the UK, 1974–2012.

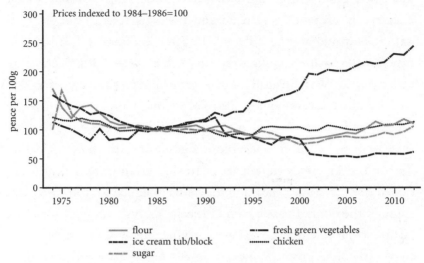

Source: Constructed from data from DEFRA. Steve Wiggins and Sharada Keats, 2015, "The Rising Cost of a Healthy Diet," Overseas Development Institute, London.

used to be for American shoppers. Energy-dense foods such as cakes and burgers have become far cheaper now in comparison to fruits and vegetables. When hovering between this and that food in the supermarket, we quite naturally compare one with another for value for money and, sadly, fruit and vegetables often look unaffordable.[49] We talk about making smart food choices, but in the brute economic terms of calories-per-buck, it might be smarter to buy a block of ice cream than to take a risk on a bag of carrots that your family may or may not eat.

In the United Kingdom, from 1997 to 2009, the price of fruit and vegetables rose 7 percent while the price of junk foods fell 15 percent. Likewise, in Brazil, China, South Korea, and Mexico, the price of fruit and vegetables rose by 2–3 percent on average each year from 1991 to 2012—twice the rate of most other foods. Meanwhile, ultra-processed foods such as packaged cakes, chocolate, snacks, and blocks of ice cream have all fallen in price.[50]

Many pundits talk about junk food as palatable food, as if we could never learn to desire a crisp green salad or a warming bowl of deep-orange pumpkin soup. Yet if we shopped in a food environment that promoted greens with two-for-one deals in the way that it currently promotes chocolate and breakfast cereal, our individual tastes would also be different. In 2011, economist Tyler Cowen experimented with shopping for a month at an Asian grocery store called Great Wall instead of the typical American supermarket where he had always bought his food before. Cowen found that he bought and enjoyed vastly more greens than in the past, because they were so much cheaper, more abundant, and more appealing than at the American supermarket. In this Asian store, greens functioned as a loss leader, meaning the items designed to lure customers through the door. The selection of leaves, shoots, and pods included "Chinese garlic chives, sweet potato vines, baby Chinese broccoli, chrysanthemum greens, snow peas, green beans, baby red amaranth, Malaman spinach, yam tips," and more, including six types of bok choy. Cowen found that the price of these delicious greens was a fraction of what he would have paid in the nearest Safeway. Green peppers were just 99 cents a pound, compared with $5.99 at the Safeway. After a month of shopping at Great Wall, Cowen found that he began to enjoy greens far more and started to choose them almost automatically.[51]

Economic food policy is often completely out of joint with current health advice. Governments lecture the poor to eat more fruits and vegetables, but they do not—by and large—make vegetables the easy option on a tight budget. Price is far from the only obstacle to eating vegetables. Poorer families are less likely to have the cooking facilities in which to whip up a pan of wilted rainbow chard. It's not easy to cook fashionable plant-based meals when the only utensils at your disposal are a kettle and a microwave. If you are doing long hours of low-paid and irregular shift work, you also may not feel you have the brain space or energy to peel carrots or slice cauliflower. James Bloodworth spent six months undercover doing low-wage work in Britain's gig economy and found that he met no one doing this kind of job who came home and stood around for half an hour cooking themselves broccoli. "When we walked through the door at midnight at the end of a shift, we kicked off our boots and collapsed onto our beds with a bag of McDonald's and a can of beer," Bloodworth writes.[52]

But it would help if greens were more affordable. There is clear evidence that price increases have an immediate impact on how many fruits and vegetables we buy, especially for those on low incomes. One study from the United States looked at long-term data from nearly four thousand children aged six to seventeen in 1998, 2000, and 2002. The researchers found that when the price of fruits and vegetables rose by as little as 10 percent, this was associated with a 0.7 percent increase in the children's body mass index.[53] For a twelve-year-old girl, this translated into a weight gain of half a pound—not a huge amount, but still significant. The price of greens matters.

The bigger problem here is that our culture still does not seem to recognize the true value of high-quality food and the difference it makes to human lives across the board. I was once having a conversation with a British head teacher about some of the children at his school and why they tended to eat such poor lunches, lacking any vegetables. Several of these children were already obese at the age of nine or ten. Some had parents who were drug addicts, and others had suffered abuse. "You don't understand," the head teacher told me. "Food is the least of their worries." He was surely right that healthier food is not the answer to everything. In this world of deprivation, heartbreak, and pain, not every problem can be salved with better diets. But unlike so many other human agonies, the pain of poor diet is an issue we can do something about.

The true value of food goes beyond price, and once we collectively start to realize this once again, the challenge will be for policy makers to build food environments that encourage people to make better food choices rather than berating them for making bad ones. Instead of subsidizing sugar and soybean oil and cheap corn, governments could subsidize the cultivation of greens. Research suggests that subsidizing vegetables by just 5 percent could increase vegetable consumption for low-income consumers by more than 3 percent.[54]

Like anything else that we buy, food is subject to the bizarre workings of market forces, but unlike other consumables, good food is essential to the quality of life—a commodity for which there is ultimately no substitute. Populations are unlikely to start making better food choices until governments start to see that money spent on regulating the quality of food for everyone is never money wasted. To invest in better-quality food is to invest in the land and air but also in health and joy. Generations to come will need to build a new vision of prosperity and of food.

I take hope from remembering that governments have not always been quite so hands-off as they are today about food standards. In eighteenth-century Paris, the quality of a basic loaf of bread was sternly policed because, as historian Steven Kaplan has written, the presence of bad bread in the market was seen as a "mark of social breakdown." If any consumer believed that he or she had been sold a loaf of bread that was of poor quality or short weight, they could take it to a police officer, who would check the loaf and, if convinced that it was defective, would call for the baker and issue a fine. Imagine if we had the same draconian food laws in place today: How many millions of substandard loaves and baps and buns from how many thousands of supermarkets would require immediate police action?[55]

Such a law could never—and almost certainly *should* never—be enacted in the modern world. But it's a useful reminder that societies have not always been so careless about maintaining the quality of food for sale. Sooner or later, maybe we will once again recognize that a prosperous life without good food is no prosperity at all. As the old saying goes, "You can't eat money."

CHAPTER FOUR

OUT OF TIME

IN 1969, A GROUP OF MEDICAL RESEARCHERS DECIDED to study what happened to the health of Japanese men when they left Japan and settled in the West. The researchers knew that the average Japanese man in Japan was much less prone to heart disease than the average middle-aged man in the United States. But what about Japanese men in America? Would they follow an American pattern of heart disease or a Japanese one?

At that time, men in the United States were already suffering some of the highest rates of coronary heart disease in the world, whereas the incidence of heart disease in Japan was unusually low, especially by the standards of rich countries. The most obvious explanation for these differing health outcomes was diet. When it comes to heart health, habitual meals of burgers, pizza, and soda pose more of a risk than a traditional Japanese diet centered on fish, vegetables, tofu, green tea, and seaweed.

Sure enough, medical data gathered in the San Francisco Bay area confirmed that the heart health of Japanese men in California was significantly worse on average than that of Japanese men in Japan (though still better than the average American male). These Japanese

Americans were more likely to suffer chest pain and heart attacks than their counterparts living in Japan. This would seem to tie in with the theory that swapping a traditional Japanese diet for a Westernized one was what gave these Japanese men their elevated risk of heart disease.

But when the researchers—led by Michael Marmot—dug deeper, they found that diet alone could not explain why so many Japanese men died of heart disease in the United States compared to their counterparts in Japan. It was not just Western food that put a strain on the heart; it was also Western culture, with its emphasis on speed and individualism.

Marmot and colleagues studied the lives and health records of nearly four thousand Japanese men over the age of thirty living in San Francisco and Oakland. Some of these men had very low rates of heart disease, comparable to Japanese men back in Japan, while others had much higher rates. When they crunched the data, they found that the difference could not be accounted for solely by the food these men ate (nor by other risk factors such as smoking). There were high rates of heart disease even among many of the Japanese Californians who reported eating a traditional Japanese diet.

More than their food itself, what mattered for the health of these men was the surrounding culture in which they ate it. The Japanese men in California filled out a questionnaire assessing how much they had assimilated to American cultural and social values. The results showed that between the most Westernized and the least Westernized groups there was a *fivefold* difference in coronary heart disease, regardless of their diet preferences. Those who led lives that were culturally the most Japanese were the most protected against heart disease. The heart health of these men was affected by factors that at first glance seem completely irrelevant, such as whether they spoke Japanese to their children, how much they had been exposed to Japanese culture during childhood, and whether they socialized with other Japanese people.[1]

Here was confirmation, if it were needed, that how we eat is as important for our health as what we eat. Eating Japanese food was not by itself enough to give these men low levels of heart disease. To get the full benefit, they also needed to slow down and eat meals in a Japanese way, re-creating the culture of their homeland in the sun of California. Marmot argued that the Japanese Americans with the lowest risk of heart disease benefited from "a stable society whose members enjoy the support of their fellows in closely knit groups," a sense of community that was not experienced by the Japanese men who had become more Americanized. In every aspect of their lives, these men benefited from the stress-reducing qualities of living in a culture organized around group cohesion and common values.

Even if they were eating the same foods, the more culturally Japanese men were organizing their meals in a different way: in less of a rush and a panic and with more of a sense of ritual. Whatever they were eating, they made time to eat it. By contrast, studies of American men around the same time found that those who were most at risk of heart disease had certain behavior patterns characterized by individualism, impatience, and a desperate sense of urgency about time, all qualities that American society strongly promoted.[2]

Stage four of the nutrition transition has not only entailed a change in what we eat. It is also an obliteration of the rituals of *how* we ate in the recent past. Our health is affected by the rhythms and rituals of eating as much as it is by the content of our diets. We have been sold the idea that all that matters about food is the nutrients it contains. But an organic salad gulped down in a state of anxiety and solitude is not necessarily a healthier meal than a takeaway of fish and chips enjoyed at leisure with friends.

Time scarcity is one of the great underexplored reasons why modern food habits differ from those of previous generations. There have been times when, after a day of work and shuttling the children here and there, I've looked at the dining supplements in the weekend papers and laughed at the ambition of someone's idea of a quick and simple family meal that involved multiple pans, rare spices, and elaborate cooking methods. It often felt as if there weren't enough minutes in the evening, nor space in my tired brain, to create the dinners I aspired to.

A lack of time—or a perceived lack—hovers over most of our modern food habits, thwarting our desires and forcing us into compromises we never quite intended. There is a gathering body of evidence that when someone feels lacking in time, he or she will cook less, enjoy meals less, and yet end up consuming more, especially of convenience food. Sliced bread was only the start. Everywhere you look there are products promising to save you time, from the soggy monstrosity that is quick-cook pasta to two-minute rice. All this talk about time is a clever marketing device, too, because it can convince us that there is no point even trying to cook anything that takes longer than twenty minutes (even though that same twenty minutes feels like nothing when we are browsing online shopping). Feeling rushed makes us buy more takeaways and use our microwaves more and our wooden spoons less.

The rhythms of life have changed, and our diets often come out the losers. Bad timing and tricky routines can thwart our best eating intentions. One American worker doing long hours of low-paid shift work told researchers that the only way she would be able to eat vegetables was if she changed jobs. As things stood, she couldn't see how she could make them a part of her life, helter-skeltering between shifts.[3]

This worker is not alone in feeling that time to eat has been squeezed out of the day. Yet there is something paradoxical in our perception that we have too little time to cook or eat properly. By absolute objective measures, most of us have far more free time on average than workers did a hundred years ago: nearly a *thousand* more hours a year, in fact. In 1900, the average American worked 2,700 hours a year. By 2015, the average American worked just 1,790 hours a year and owned a kitchen containing whizzy time-saving gadgets his or her ancestors could only dream of.[4]

When we say we lack time to cook—or even time to eat—we are not making a simple statement of fact. We are talking about cultural values and the way that our society dictates that our days should be carved up. The changing rhythms of life have affected our eating in some profound and surprising ways. A sense of time pressure leads us to eat different foods and to eat them in new ways. A collective obsession with not wasting time has contributed to the rise of the snack and the fall of the cooked breakfast, to a rise in convenience foods and the death of the lunch hour.

"I think it's about priorities," says a woman from Trinidad whom I started chatting with one afternoon. She works full time at a high-powered job and has three kids who do intensive levels of sport after school, yet she says she always makes sure there is a home-cooked dinner on the table, even if it is a dish of reheated leftovers. Dinner for her is nonnegotiable, just as it had been growing up in Trinidad, and she feels dismayed that so many people in modern Britain do not seem to feel the way that she does about the importance of arranging the evening around the preparation and eating of food.

I half-agreed with her, but I also wanted to say that our choices about food are not made in a state of perfect freedom. The really dismaying aspect of our modern food culture is that so many people have been forced by the circumstances of their lives to make time for meals a lower priority than they would like. Time is one thing and timing is another. We may have more free minutes of time than in the past, but all too often, these minutes do not fall at mealtimes, when we actually need them. The coronary-inducing culture of individualism and impatience that Michael Marmot observed among Japanese American men in California in the 1970s has now spread across much of the world.

THE DEATH OF THE LUNCH HOUR

Being a female textile worker in 1920s Westphalia wasn't exactly a life of leisure. Westphalia was an area of northwest Germany famous for both its ham and its cloth. In 1927, a researcher called Lydia Lueb met two thousand young Westphalian women working in the cotton and flax industry. Lueb asked these spinners, weavers, and skilled needle-women how they spent their free time (or *Freizeit*). But the truth was that between the demands of work and home, these young women on modest incomes hardly had any time at all that was free in our sense of the word. They never ate out and most had not traveled beyond their own small village or town. When asked what their single favorite activity was, by far the most popular answer (given by 41 percent of the sample) was simply "rest."[5]

These Westphalian women devoted on average fifty-four hours a week—including Saturday mornings—to slaving away over pieces of cloth. They had just one week's holiday per year, and half of them spent it—once again—on needlework or in the garden. Even their Sundays were full of duties: church in the morning, then housework and lunch, and after that family visits.

In one respect, however, these hardworking women had some-thing that has become an almost unheard-of luxury in our own society. Each day at work, they stopped for a lunch break of a full seventy-five to ninety minutes. These breaks were not staggered but synchronized, running from 12:00 p.m. to 1:15 p.m. or 12:00 p.m. to 1:30 p.m., depending on the factory. When it was lunchtime, every single woman stopped work in unison to eat. Industrious as these Westphalians were, there was no question of them feeling too busy to feed themselves, because scheduled time to eat was factored into their days. A long lunch break was woven into their routines as securely as flax into linen. They never lacked the time to eat.[6]

Compared to those poor female Westphalian textile workers in the 1920s, the average worker today is swimming in free time. Except, it seems, for time to eat.

When we say we are lacking in the time to eat well, what we often mean is that we lack synchronized time to eat. Our days and weeks are broken up with constant interruptions, and meals are no longer taken communally and in unison but are a cacophony of individual collations snatched here and there, with no company but the voices in our headphones. Many of us, to our own annoyance, are trapped in routines in which eating well seems all but impossible. Yet this is partly because we live in a world that places a higher premium on time than it does on food.

A sign of how little we value eating is that lunch breaks have stopped being a normal and expected part of the working day. Rich or poor, most people used to be able to count on having time to eat (assuming you had the money to buy something to satisfy your hunger). Now, all over the world, the working lunch is getting squeezed out of the day by the fast pace of life. FORTY-FIVE MINUTES IS THE NEW HOUR announced a poster I saw the other day in Farringdon, near the banking district of London, advertising a gym where busy workers could use their lunch break to dash through a "super-concentrated workout." The lunch break, if it exists at all, is often used for other, supposedly more important activities, such as shopping or exercising or just more work.

It's not that long, structured lunch hours of the kind enjoyed by those Westphalian workers were always seen as normal. Ideas about the right hour to eat dinner have changed many times before. We sometimes beat ourselves up over our inability to adhere to a 1950s pattern of three square meals, forgetting that this way of eating was not always the ideal. There have been plenty of times and places where meals were informal, snacky, haphazard affairs, eaten without much ceremony or cutlery. Before the Industrial Revolution, fieldworkers might stop and unwrap a hunk of bread and cheese whenever their hunger or a pause in the work dictated. It was only with the huge influx of workers from fields to factories in the nineteenth century that the idea of fixing the lunch hour by the clock became formalized in Europe and the United States.[7]

Yet the changes to mealtimes that we are experiencing nowadays are something else altogether. Different cultures have always slotted eating occasions into the day at different moments. But never

Figure 4.1.   How time is used during the day.

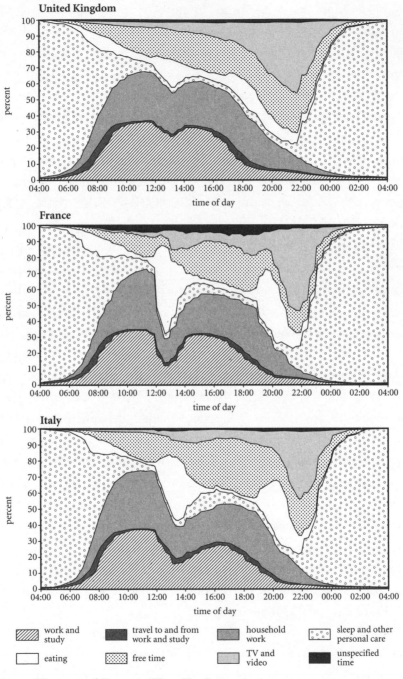

Source: Harmonised European Time Use Survey.

before have so many populations around the world organized life in such a way that shared time to eat is more or less scheduled out of existence.

If you are of a nosy disposition, as I am, there are worse ways to while away an afternoon than by scrolling through graphs showing how different Europeans spend their time. From around 1998 to 2006, researchers in Europe gathered data on time use for fifteen different countries.[8] Thousands of people (more than twenty thousand individuals in Italy, nearly four thousand in Sweden, and so on) were asked to keep diaries recording how they spent their days. The data was then compiled into a series of tables—*areagrafs*—showing how the hours of the day were portioned up in different European countries around the year 2006.[9]

What makes these time-use statistics so fascinating is that— imperfect as such data necessarily is—you seem to be peering through a secret window revealing the truth about human behavior across different countries. Everyone may sleep and eat and work and rest, but different people slice up these activities in different ways. The average Bulgarian, for reasons that are not clear to me, sleeps more than the average Norwegian.

On these graphs, the human day is reduced to a series of crude activities mapped across twenty-four hours in different colors. The graphs start and finish at 4:00 a.m. The different colors show what percentage of the population is engaged in a particular activity at any given time. At 4:00 a.m., for example, you can see that virtually all of Europe is asleep (or trying to sleep), a fact represented on the graphs in thick swaths of blue. From 6:00 a.m. to 8:00 p.m., there is a chunky red splotch of time that represents work or study.

Eating is a special case on these time-use graphs. The pattern of time devoted to food varies from country to country far more than work time or rest time. The hours spent eating are colored white and make quite different shapes in the pattern of the day depending on the country. In the graphs for France, Spain, Bulgaria, and

Italy, eating occupies clear and distinct peaks of time—deep bulges of white—that push work and rest away for a while. These peaks are smallest at breakfast time. Even in France, land of croissants and café au lait, there is only a mild bulge of eating time in the morning—a slight increase in the white section of the graph from 6:00 a.m. to 8:00 a.m.—suggesting that breakfast is considered an optional meal by many. Lunch is another story. The white part of the graph dramatically rises from 12:00 midday to 2:00 p.m. in France and Italy and from 1:30 to 4:00 p.m. in Spain. These graphs suggest that lunch, for most French, Spanish, and Italian people, still happens at a regular and predictable moment in the day. There is a second big bulge in eating time in the evening, from 7:00 to 9:00 p.m. in France and 9:00 to 11:00 p.m. in Spain. These graphs reflect a way of life in which there is still agreement on when lunchtime or dinnertime should be.

In France, Italy, and Spain, most of the population in 2006 still ate to a common beat. In other places, this old tempo had already become radically disrupted. To turn from the Spanish graph to those for the United Kingdom, Poland, Slovenia, Sweden, and Norway is startling. In these countries, the white band of eating time had become something with no clearly defined bulges but rather a continuous ribbon throughout the day. Eating was something that equal numbers of people might be doing at any point from 6:00 a.m. to 10:00 p.m. In Norway and the United Kingdom, there were slight increases in the percentage of the population eating around 12:00 to 1:00 p.m.—some remnants of the old lunch hour—but other than that, eating was something that around 10 percent of people, give or take, might be doing at any given time. This is not to say that no one in these countries had fixed mealtimes. In an individual Norwegian household, dinner might have been timed each day to start strictly at 7:00 p.m. But judging from the data, there is no particular likelihood that other families would be eating at the same time.

The *areagrafs* offer a stark illustration of societies that have lost any communal sense of mealtimes. In Poland or Sweden, you are as likely to be eating at 4:00 p.m. as at 8:00 p.m. For millions of people, there is simply no normal pattern for mealtimes anymore.

Meals are not just a way to use up time but a series of ceremonies through which we experience time. Like religious worship, or news on the radio, eating used to punctuate the day at certain set moments. Even if you were eating lunch alone, you knew that much of the country was doing the same thing at that exact moment, and this imbued your solitary meal with a particular rhythm. You were doing the right thing at the right time. Now, our eating is often out of sync. You may go to a café and order all-day breakfast at nine o'clock at night or buy an ice cream along with the morning paper without anyone looking at you askance. In many places, time for eating is now a thin ribbon that runs throughout the day, often with little coordination even between people living in the same household.

In the years since this data was assembled, the shared mealtime has become still more disrupted. Even in Spain and Italy, time for eating is now becoming something shorter and more matter-of-fact. After the Great Recession of 2008, many businesses in Spain cut the traditional two-hour lunch break to a single hour. In France, once a bastion of slow meals, Parisian workers have started consuming such ad hoc novelties as ready-to-eat salads and quick repasts of bulgur wheat salad or sandwiches from snack bars rather than multicourse lunches at a brasserie.

Today, organizing two or more people with different tastes and schedules to eat together can seem like a feat of engineering. The individualism of modern life pushes forcefully against the shared meal. If you want to eat together, you have to synchronize more than just the timing of the food, though that can be tricky enough. (Many is the dinner when a pan of rice has been all cooked and beautifully fluffed up, only for the cook to realize that the curry to go with it won't be ready for another twenty minutes.) A meal is something that orchestrates people as much as food. The traditional fixed family dinnertime makes certain assumptions: that people in a family are all free to eat at the same time in the same place and that they share a common appetite for the foods on the table. In our world, these are big assumptions to make.

In 2009–2011, a team of London researchers set out to examine the conventional wisdom that modern families were time pressed by interviewing forty dual-earner households about how they organized their meals. A common theme was that, much as parents valued the idea of family meals, it was often very difficult to synchronize everyone to eat at the same time. Most of the families interviewed aspired to eat together, but for one reason or another, fewer than a third of them managed a shared dinner around a table on most weekdays.

Family meals eaten apart are not necessarily cause for alarm. There can be a kind of moral panic about the decline of family Dinner with a capital *D*, but meals can be enjoyed and shared in many ways, even when you are not all lined up breaking bread like at the Last Supper. I often think my favorite meal of the whole week is the supper I eat on Sunday night, after a yoga class, when everyone else has left the table and I am free to sit alone with a salad or a bowl of noodle soup and a book. Bliss.

But something important about eating is lost when meals are never—or almost never—timed to be taken together. There's an old word, *commensality*, which literally means eating at the same table. The food anthropologist Claude Fischler has written that commensality is what provides the fundamental human script of eating in every society. It was how basic bonds of kinship were forged.

It is no small change to our eating habits when families lose this commensality. For many of us now, meals have become a question of ingesting nutrients rather than of sharing flavors or spending time together. In one London family of five, the mother told researchers that the only time everyone ate the same thing at the same time was on Christmas Day.[10] The rest of the year, it seemed impossible to synchronize routines and tastes. The mother and father (who worked as a debt collector with unpredictable hours) were on two separate weight-loss diets, while their younger daughter, aged eleven, was an extremely picky eater who only liked foods such as ready-made pizza and fries, not allowed on either of the parents' diets. Meanwhile, an older sister, aged twenty-one, who lived at home with her boyfriend, usually bought her own food—things like takeaway kebabs. On a typical evening, the mother served her eleven-year-old a meal of convenience food at 4:15, then cooked herself a diet meal of steak and ate it with cottage cheese and tomatoes, and finally cooked another meal of chicken breast and salad for her husband when he came in from the gym.[11]

This is a description of a household for whom eating has become something totally uncoordinated and unshared. The family members are separated from each other both in what they eat and when they eat it. It's not that this mother is failing to spend time on cooking. On the contrary, she is putting great thought and energy into preparing food for herself, her daughter, and her husband. Each day, she is cooking for an hour and three-quarters, from 4:15 to 6:00 p.m., trying to meet everyone's needs including her own. She is probably spending far *more* time than she would if her family ate a single meal at a single time. But differing preferences and routines somehow stop them from sharing a single block of eating time.

Our loss of shared eating time has consequences. This is not a plea to turn back the clock to a patriarchal dinner table where a mother cooks and a father keeps order. It's about holding on to the principle that time to enjoy food remains a basic human need, even as modern families evolve. As the example of the Japanese American men shows, a culture that denies people the time to pause over meals—at least from time to time—can end up damaging our health. All too often, what looks like a problem of poor diet is at root a problem of hurried routines reinforced by a culture of extreme individualism.

The individualism of modern eating affects both the way we feed ourselves and the way we are fed by others. When it comes to feeding ourselves, many have come to feel that we never have to eat so much as a morsel that we don't want to eat. We may choose to be gluten-free today but not tomorrow, just because we can. The other day, I saw a young woman having a meal in an independent café. To go with her spiralized salad, she had brought in her own cup of Starbucks latte.

In the near-infinite array of modern food choices, we can be as fickle as we like, and the idea of a fixed mealtime with a single shared main dish can seem like an absurd imposition, as outmoded as a world in which there were only three channels on TV. The breakdown of mealtimes is part of a larger social fracturing, in which many people no longer feel obliged to conform to the values of the whole table.

But this erosion of social obligations cuts both ways. Fixed mealtimes used to operate as a kind of contract between eaters and feeders. Both parts of that contract have now disintegrated. When I was a child, I remember feeling a duty to sit at the table and eat what I was given at a set time, but this was partly because someone had taken the trouble to feed me. Now, especially in the world of work, it can feel as if no one really cares whether we eat or not.

LIKE A REST IN MUSIC

When Anne Marie Rafferty was a student nurse in Scotland in the late 1970s, she remembers, lunch breaks were a highlight of the day. In common with all the other nurses, Rafferty worked eight-hour shifts with a break in the middle, and everyone got the full time allotted. There was a "set social hierarchy" for dining, with separate dining rooms for men and women. The food was wholesome and cheap enough, thanks to subsidies, that everyone sat and ate it together. Nurses the world over spend their working life looking after others, and mealtimes used to be a rare moment when their employer looked after them for a change. "The lunch break was like a rest in music," Rafferty recalls. "It was there to break up the rhythms and revive you and sustain you for what was next."

In those days, the staff cafeteria would serve hearty two-course lunches to hospital staff. Rafferty remembers relishing the food on offer: "institutional comfort food type stuff" like roast beef or chicken, lots of stews, always plenty of vegetables with hot puddings and lashings of custard or a starter of warming soup if you preferred. Rafferty recalls that "absolutely no one" in that hospital bought food to take away from the cafeteria. There was plenty of time to eat the full meal in the cafeteria and maybe relax with a cup of coffee and a cigarette afterward ("in those days, we all smoked in hospitals").

Those ways of eating have gone, and not just in Britain. At the age of fifty-nine, Rafferty is energetic, cycling round London wearing a bright-orange cycle helmet. She now works as a professor of nursing at King's College London and is saddened to see how nurses today are denied the leisurely lunches that she once enjoyed. In the 1980s, Rafferty noticed that commercial "snacky stuff" started to appear both on the wards and in the staff dining areas. Vending machines crammed with chips, chocolate, and fizzy drinks started to materialize. Hospitals were rebuilt on a much larger scale, and the old staff dining rooms and kitchens were often eliminated.

For many nurses now—around the world—poor eating habits are almost built into the structure of the day. The life of a nurse is much altered now from how it was in the 1970s. No one smokes in hospitals anymore, which is progress, but in other respects, it's much harder for nurses to maintain a good state of health at work than it was for earlier generations. Twelve-hour shifts have become common in place of the old eight-hour ones, yet even as the working day (or night) has become longer, eating breaks have shrunk.

It has often been remarked what an irony it is that healthcare professionals—the very people who are trying to help the rest of us become more healthy—are actually more likely to be overweight or obese than the general population. How is it that healthy eating is so elusive for people who know so much about health? But this is looking at the problem the wrong way around. Nurses may be richer in health knowledge than the rest of us, but the thing that they are poorer in is time to eat.

As of 2008, the average body mass index of an American nurse was 27.2, somewhere between overweight and obese. The conclusion sometimes drawn—by cruel commentators—is that obese nurses must not be very good at their jobs since they set their patients such a bad example. But the poor diet of nurses now is generally not caused by their *failure* to do their jobs but because they work all too diligently, following routines and schedules that make eating decent meals all but impossible. For too many nurses, the very timings and pressures of the working day make it likely that the whole experience of eating will not be a healthy one. In 2017, I interviewed Kerry Hart, a young healthcare worker in the United Kingdom who said that all the nurses in the unit where she worked—a foot clinic in a big research hospital—struggled to eat regular meals, in part because they put the patients' needs before their own. Hart told me that she and the other nurses would often cut short their own lunch break to spend a few more minutes with a patient and then find that there was hardly time to heat up a bowl of microwave soup before they were due back on the ward.[12]

Imagine you are a nurse in the middle of the night, nine hours into a twelve-hour shift, looking after sick people without a break and drowning in paperwork. Your body is crying out with tiredness and hunger. You feel in need of comfort and something to wake you up. And just down the corridor is a vending machine selling sweet caffeinated beverages, puffy potato snacks, and chocolate bars. You know that there's nowhere in the hospital open to sell hot food, and you also know that even if there were, the manager wouldn't give you half an hour off because the feeding of employees is not one of his or her targets.

Night-shift workers are eating under double time pressure. Like so many others, they feel lacking in the sheer minutes and hours needed to prepare and eat anything wholesome. But they also suffer from bad timing, with their work forcing them to ingest food at the wrong time. Across the world, night-shift workers are prey to higher levels of diet-related health problems, ranging from heart disease to type 2 diabetes and obesity, more so than daytime workers.[13] Staying awake to work overnight messes with the circadian biological clock, which has a knock-on effect on metabolism. Usually the body is in a fasting state overnight, so to eat at a time when the body expects to be resting can create an exaggerated glucose response compared to the same meal eaten during the day.

To add to the woes of the night-shift worker, there's also the fact that the food choices available in the middle of the night tend to be very limited and ultra-processed. A group of Australian firefighters in Melbourne told researchers in 2017 that they would often binge on chocolates and sweet biscuits after a night call because there was nothing else in the station to eat. Compared to firefighters on day shifts, those on night shifts were more likely to eat sugary or salty snack foods. There was a spirit of camaraderie among the men; they often made a group decision to head to McDonald's or get a pizza in the middle of the night, and it wasn't easy to opt out.[14]

Neither nurses nor firefighters are lacking in willpower, but like many workers nowadays they may feel almost forced into unhealthy meals by the timing of the day and their workplace environment. This applies to children as well as adults. One study of schools in Wales found that allocating even a few minutes extra for a lunch break could make the difference between whether children selected fruit and vegetables or not as part of their school meal. The shorter the lunch break, the more likely they were to eat fries and the less likely to eat vegetables.[15]

When did we start to see meal breaks as a disruption of our working lives rather than something that patterns our days and gives meaning to them? The squeezing of meals out of the working day is part of a larger shift—exacerbated by email and smartphones—in which people feel they must be on duty at all times. The majority of workers in America do not take their full allotment of holiday time, scared of falling behind or of being perceived as less dedicated than their coworkers. Many modern teenagers also seem to be constantly on edge for the latest notifications from their phones, unable to give themselves up fully to the eye contact and slower pace of a communal meal.

How different life—and eating—would be if we had the lunch hours of the past and the culture that underpinned them. Most modern employees would not wish to change places with a Westphalian textile worker in the 1920s. We would not enjoy the rigid hours, the lack of freedom, the sheer monotonous grind, or the compulsory Saturday-morning shifts. But oh! for the ninety-minute lunch break.

WASTING TIME OR WASTING FOOD

"My mother would have a heart attack if she saw you trimming beans like that!" hissed cookery teacher Nikita Gulhane one April afternoon when he asked me to top and tail some green beans to add to a potato and coconut curry. I was in a friend's kitchen, and Gulhane was teaching us how to make a series of authentic Indian dishes. I had tackled the beans in my usual impatient fashion, holding them in a bundle and slicing through multiple beans at once, ignoring the fact that this left behind trimmings of variable lengths. I know it's not perfect, but this is how I trim beans at home when I feel I need to get them in the pot in two minutes flat. Which, to be frank, is almost every time I cook them.

We had been getting on so well up till then, but Gulhane's displeasure at my slovenly ways was palpable. He was brought up in Britain by an Indian mother—he calls her "Mrs. G"—who taught him to pare beans (and all vegetables) down to the very tips, removing only hard woody stalks that were truly inedible. If the task took a little longer, so be it. Gulhane now lives in north London, where there are few taboos on trimming vegetables in an overlavish fashion, yet he still cannot bear to discard so much as a cauliflower stalk. Mrs. G—who came from the western region of India, not far from Mumbai—taught him to go to any lengths to avoid wasting food.

My squabble with Gulhane over green beans shook me, because I realized afterward—to my great annoyance—that he was right. I started to think differently about the relationship between food waste and time. Once, cooks and eaters the world over—like Mrs. G— tended to be people who hated to waste food. Much cuisine was the art of salvage: an often intricate and time-consuming business of saving scraps that might otherwise rot or go to waste. In the nineteenth century, cooks would keep a grease pot near the stove, made up of the white grease left over from frying bacon or salt pork, for reuse.[16] This degree of frugality is almost unimaginable today. How many families now would look at the white grease left over from frying a pan of bacon for breakfast and see it as nourishment rather than something to discard as quickly as possible?

Once we hated to waste food; now we hate to waste time, which has become the "ultimate scarce commodity," as John P. Robinson and Geoffrey Godbey, experts on American time use, wrote in 1997.[17] The consequences of this change are all around us. Research in 2004 from WRAP (the Waste and Resources Action Programme) showed that the average person in the UK wasted food worth £424 a year. More than half of UK households threw away whole pints of milk, whole loaves of bread, cheese, cooked meat, and half bottles of wine all because there was no chance to consume them. These patterns of wastefulness are mirrored in every developed country in the world (not helped by the "use by" and "best before" dates that encourage us to discard tons of food that is still perfectly edible).

Food waste has many causes and takes many forms. In developing countries such as India, most of the waste happens at the production stage, in the fields. As much as 40 percent of all fresh food in India rots before it reaches the market because of faulty transport and storage. In rich countries, by contrast, the greatest problem is consumer waste. Sometimes all this food waste is painted as a moral failing on the part of a younger generation, but like other aspects of the way we eat, waste is environmental and structural. Much of our food waste is driven by the way our food is supplied to us by retailers, who encourage us to buy more food than we can eat with two-for-one offers and who keep their shelves continuously full.[18]

Food waste—in its many guises—is also a byproduct of a culture that makes us feel rushed off our feet. Feeling short of time at the shopping stage makes us buy more food than we need, as well as purchase such time-savers as quickly cooked and expensive chicken breasts rather than getting a whole chicken, jointing it, and using every last scrap. At the cooking and eating stage, time pressure makes us scan our fridges too hastily, ignoring things that need to be eaten up before they go bad.

The economist Gary Becker (1930–2014) first articulated, in 1965, the idea that Americans had changed from people who economized on food into people who were frugal with time.[19] Becker's article "A Theory of the Allocation of Time" was a groundbreaking essay that helped established a novel branch of economics: New Home Economics. Becker and his colleague at Columbia University, Jacob Mincer, were interested in explaining human behavior in terms of domestic households rather than seeing our actions as purely individual.

Becker noticed that in recent years, his fellow Americans had entered into a new relationship with time. They had more of it yet hoarded it more zealously. There had been a "large secular decline" in the number of weeks people worked, Becker found, which meant that in most countries workers were only at work for around a third of the time available. Meanwhile, 1960s consumers had even more free time than their prewar forebears thanks to technologies that helped them maximize their nonworking time: cars, electric razors, telephones, and more.[20]

Yet, oddly, possessing all this new free time only seemed to make people feel more harried. Becker found that Americans of the 1960s were more conscious of time than ever: "they keep track of it continuously, make (and keep) appointments for specific minutes, rush about more, cook steak and chops rather than time-consuming stews and so forth." What was the cause of this American obsession with time? The key fact for Becker was that the market value of time relative to the price of goods (including food) had become higher in the United States than it was elsewhere, particularly for women.[21]

Partly what has changed is the way that time in households is divided up between the sexes. Since the 1930s, men of working age have done on average less paid work, but women have done a great deal more. American women did eleven more hours of paid work per week in the year 2000 than they did in 1970. This colossal social change was bound to have knock-on effects on the rest of life, leaving both men and women hard-pressed to find time to cook. Women felt harried because they were trying to juggle both paid work and housework. Men, too, felt they lacked time because, although they were working fewer hours, they were suddenly expected to do more housework. Meanwhile, both sexes devoted far more time to childcare than in the past, which left even fewer hours for cooking (although the kitchen, in my experience, can be the best place to keep children amused).[22]

Until recently, preparing food occupied a certain nonnegotiable number of hours in the day—for women, anyway. Over the past two or three decades, by contrast, making food has become something that we can choose to spend fewer of our precious hours on. As of 2001, 64 percent of American men and 35 percent of women ages twenty-one to sixty-four reported to the American Time Use Study that they spent no time at all on daily food preparation.[23]

When we lament the decline of time spent on cooking, we need to be clear what it is that we are lamenting. Many of the female cooks who devoted so many hours to preparing food in the past did so because they did not think their own time was worth much. A mother of the 1950s, especially if she did no paid work, might devote time to browsing food markets for the cheapest seasonal produce and clip coupons to save a few pennies here and there. By contrast, noted Becker, when a woman's time is valued more highly in market terms, she might find paid work and spend less time cooking by relying on convenience meals. This was not a sign to Becker that American mothers were becoming lazier. It was usually a rational economic decision. Becker's big idea was that if economists really wanted to understand the income of a household, they should look not just at a family's money budgets but its time budgets: how much free time different people in the family had and—just as important—how they valued that time.[24]

There are still those in the modern world who place a high value on food. Even in a time-poor economy, it should still be perfectly possible to cook and eat a quick meal that does not feel like a race against the clock, but you may have to adjust your ideas of cooking to fit the time available. If you are in Spain, perhaps you take a piece of bread, toast it, rub it with pungent garlic and salt, pour over good olive oil and pulped tomatoes, and relish a five-minute feast of *pan con tomate*. Such a meal is quickly made and quickly eaten, but that is how it is meant to be.

Instead of lamenting the loss of the dinners only a 1950s housewife would want to prepare, we need to adapt our food to the new rhythms of life. Ours is not the first generation to worry that eating was getting too fast. We could learn something from Edouard de Pomiane, a writer who was already thinking about how to eat well during busy times the best part of a century ago.

THE RHYTHM OF MODERN LIFE

From today's perspective, the 1930s looks like a decade when life was less hurried: when it was still normal to eat a cooked breakfast and read a physical newspaper, a time of hats and gramophone records. Yet from the perspective of people who grew up the century before, these were already hectic times.

At the age of fifty-five, in 1930, the Polish French dietitian and scientist Edouard de Pomiane felt that the pace of life in France had changed since his youth in the nineteenth century. Pomiane observed that French people did not always linger over meals as they once had. To Pomiane, this was not necessarily a disaster but a sign that diners should adopt new, speedier dishes to fit with the new times. "Life nowadays," he wrote, "has transformed all rhythms, and one frequently meets gourmets who have to be satisfied with food which has been rapidly cooked and, to their regret, rapidly eaten. But they are none the less gourmets for that."

In 1930, Pomiane published *Cooking in Ten Minutes*, one of the great cult cookbooks. Its subtitle is "Adapting to the Rhythm of Modern Life." Pomiane showed his readers that with the right attitude and planning, it was perfectly possible to make and eat delicious meals in as little as ten minutes (or a bit longer if you wanted to enjoy more than one dish). He aimed his book at "the student, [the seamstress or salesgirl] . . . the artist . . . lazy people, poets, men of action, dreamers and scientists . . . everyone who has only an hour for lunch or dinner and yet wants half an hour of peace." One of Pomiane's top tips was to put the water on to boil as soon as you get home. Another tip was that when frying something, you should not "wait to take your hat off before putting the pan on the fire."

Pomiane's idea of ten-minute meals is astonishingly ambitious by today's standards. As well as numerous egg dishes ("scrambled eggs with green peas are a delightful spectacle"), he suggests that the ten-minute cook may enjoy moules marinière; or hot shrimp, boiled for five minutes and eaten with fresh bread and butter; or breaded veal, fried herring, or quail en cocotte. Surprisingly, the one thing he says cannot be easily cooked in ten minutes is most fresh vegetables (except for a few things such as buttered spinach)—he prefers to rely on tasty preserved alternatives such as canned peas, precooked beetroot, and jars of sauerkraut or else to slice up some raw tomatoes. Pomiane has made me feel better about my wasteful method of trimming green beans because he says that when peeling pumpkin for fast cooking you should "cut off the peel generously, without trying to be economical" before cutting it into cubes and frying them in butter. In Pomiane's book, it is better to waste a little peel than to feel you have no time to eat fresh pumpkin at all.[25]

Pomiane's idea of speedy eating was very different from what most people today would think of as fast food. For Pomiane, the main purpose of cooking things fast was to reclaim some leisure in the evening. Having enjoyed dinner, he tells his readers to "fill a warm cup with coffee. Sink into your comfortable armchair; put your feet on a chair. Light a cigarette, send a puff of smoke slowly up to the ceiling. Close your eyes. Dream of the second puff, of the second sip. You are fortunate. At the same time your gramophone is singing very softly a tango or rhumba."

To read Pomiane is to be reminded that what is often lost today is not the time to eat per se so much as the sense that we are entitled to sit and enjoy our meals, relishing every bite, no matter how long or short the time available for them. For Pomiane, a simply cooked meal of delicious food—and wine, and coffee—was a way to enjoy what little free time an individual might have. Today, by contrast, many people see free time as the thing you only get to after you are done with eating. The logical next step is to treat meals themselves with less respect. If all you are interested in is getting calories into your body as quickly as possible, then sitting down with a knife and fork can look like a waste of time.

By the standards of a microwave dinner or a sandwich, Pomiane's ten-minute cookery is impossibly slow. Our culture of impatience is one reason, if far from the only one, why we eat so many snacks now.

THE WOMAN WHO NEVER SNACKS

"The buffet's closed." These were the words that Olia Hercules's Siberian grandmother would use in the late 1980s when she wanted to announce that supper was over. It was her way of telling Olia and her older brother that there was nothing more to eat that day.

When she was growing up in a small town in rural Ukraine in the 1980s and early 1990s, no one ever gave Olia Hercules (who was born in 1984) an after-school snack. There was no need. When the whole family—Olia, her older brother, her father, and her mother—arrived home at around two o'clock every afternoon, her mother cooked a big late lunch for them all. Often it might be something like a casserole of borscht—a meaty soup of beef, beetroot, and tomatoes—with hunks of bread. Every day, the family laid out a tablecloth and sat in the dining room together and talked as they ate. This ritual wasn't a special occasion, just a normal, everyday thing. "Life wasn't as busy and crazy then," Olia tells me one winter morning in a Middle Eastern London restaurant as we share breakfast. She has eggs with eggplant, tahini, and flatbread, with sweet cardamom-scented coffee to drink.

Back in Ukraine, where she lived until the age of twelve, it wasn't difficult to avoid snacking because the way that Olia and her family ate was, give or take, how every single person they knew in their small town ate in the 1980s: hot hearty meals, with little or nothing in between. Did Olia never pester her mother for snacks, I ask her. She shakes her head incredulously. "If you were hungry, you had a meal."

Eating patterns change fast, and habits that were taken for granted in living memory can suddenly become so unusual as to seem almost peculiar. Olia Hercules, who works as a food writer and chef in London, has become a holdout simply by eating the way she always has. In a world of grazing and nibbling and chips that come in legion flavors and textures, this is the woman who never snacks.

I got to know Olia in 2015 after the publication of her first cookbook, *Mamushka*, a collection of the comforting Ukrainian recipes she grew up with, from sour cream pancakes to roast goose with noodles and potatoes with fatty salt pork and many types of homemade pickles. Olia and her book were a wonderful mystery to me. How could such an elfin and slender person enjoy such a hearty and substantial cuisine? Just before Christmas that year, Olia invited me to a dinner at her house. She had made warm, mulled alcoholic cider and rich meatballs with roasted wedges of spiced pumpkin. Over a dessert of sweet, puffy Ukrainian bread, she casually mentioned that she never ate snacks. The three British women at the dinner—including me—were agog. No snacks ever? What, not even an idle bag of tortilla chips and guacamole in front of the TV on a Saturday night? Never a post-gym rice cake topped with dark chocolate? Not really, no, Olia replied.

After that dinner, I found myself thinking about snacking. I started wondering what modern life—never mind modern eating—would look like if you took all the snacks away. I suddenly noticed just how many extra nibbles had crept into my family's routine without me exactly planning them. Clearly, I lack the authority of a Siberian grandmother because my children almost never seem to believe that "the buffet is closed."

Part of me wished I could adopt Olia's hard line on snacks, but I wasn't sure I had what it took. It is one thing to eat no snacks when that is the expected, everyday thing to do, but something else altogether to re-create the same restraint at a time and place where snacking between meals has become normal. Wasn't there once a sense of social disapproval attached to eating between meals? If so, it has long since vanished.

Humans have never snacked quite as we snack today. It is a change both of quantity and kind. According to Datamonitor, a company that analyzes food sales around the world, snacking now accounts for half of all "eating occasions" in the United States. We eat vastly more snacks than people did a generation ago, and we also consume items for these snacks that our ancestors could never have foreseen or fathomed, such as chocolate-covered pretzels, Japanese nibbles flavored with wasabi, and peculiar energy balls made from dates and nuts that claim to be healthy even though they are sweeter than a brownie. If we no longer tell our children not to eat between meals, it may be partly because "between meals" now takes up so much of our eating lives. Snacks have been both cause and consequence of huge changes to the rhythm of the day.

In these hectic times, it's becoming ever rarer to eat like a 1980s family in small-town Ukraine—even in Ukraine itself. Ukraine is the fourth-largest potato producer in the world, after China, India, and Russia. Traditionally, those potatoes were cooked and eaten at home—boiled, fried, stewed with other vegetables, or made into potato pancakes, with a generous dollop of sour cream on the side. But now, Ukrainians—like people the world over—have discovered potato chips.

A few years after Olia Hercules left the country, in 2002, Flint Brand snack foods was established in Ukraine. Flint markets a range of chips, crackers, and "youth snacks" flavored bizarrely with everything from kebabs to veal, crab, red caviar, and "hunting sausage." The flavors of these garishly packaged Ukrainian savory snacks—which are sold throughout the former Soviet states—gesture poignantly at the hearty meals they have displaced. Except on Christmas Eve, few Ukrainians today may sit down to a traditional dinner of carp in aspic with horseradish. But a Ukrainian can buy a packet of aspic-and-horseradish-flavored bread snacks to take its place, like a salty ghost of the meals their grandparents ate.

By definition, a snack is a small thing, easy to overlook. The *Oxford English Dictionary* defines a snack as "a mere bite or morsel of food." Those morsels add up, however. The increasing frequency of snacking since the 1970s—which is a global phenomenon—means that many people are scarcely acquainted with the feeling of hunger anymore. Instead of sociable structured meals of breakfast, lunch, and dinner, the new pattern is a series of solitary snacks that we often hardly notice or enjoy as they pass through our gullet.

Without snacks—and sugary drinks, which are effectively snacks in drinkable form—we would be eating many fewer calories than people did in the 1970s. I was staggered to learn that around a third of all calories consumed by the average American adult are now made up of snacks, amounting to more than six hundred daily calories for men and around five hundred calories for women. These figures are based on self-reported data, so the true figures will be even higher. Where once there were taboos on eating in the street, such behavior has become normal. Even those of us who think we don't snack much might mark the morning with a caffè latte and a biscotti and the afternoon with a protein bar. The food industry encourages us to feed every passing hunger with a bizarre range of snack products that earlier generations could never have imagined, let alone eaten.[26]

To be a nonsnacker in the modern world has become so exceptional that it may provoke incredulity in others. Olia Hercules now lives in north London with her five-year-old son, Sasha. Every weekday, after school, he asks his mom for a snack. He sees the other parents with their after-school stashes of biscuits and croissants and crackers and fruit leather, and he wants some. Every time, she refuses. "I keep replying to him that snacking is not in our culture; it's not what we do."

It's worth noting that there are some food cultures that have never shared the disdain for snacks that characterized Olia Hercules's childhood in Ukraine. The word *snack* can mean two utterly different things. A snack can be a noun or a verb. On the one hand, a snack can refer to a particular type of commercial snack food—ultra-processed and high in sugar, fat, and salt. But a snack can also refer to a way of eating: a pattern of five or six modest-sized meals (as opposed to the two or three big ones Olia Hercules grew up with in Ukraine), which may be perfectly nutritious. In some parts of the world, small eating events are considered so important, they are dignified with their own names and scheduled like other meals. The French have a meal called *le goûter*—literally, a "taste"—which happens in the afternoon, after work or school. French children may mark the end of the day with a quick reviving snack of baguette and dark chocolate or fruit and a glass of milk. In the Spanish-speaking world, they have *merienda*, a light meal eaten either between breakfast and lunch or between lunch and dinner. This could be a few slices of *jamón* and bread, a little hunk of cold tortilla and a cup of black coffee, or maybe some toast and jam.

One country where snacks are not viewed as immoral or nutritionally unsound is India. "No Indian mother ever taught her child not to eat between meals," says the British Asian food writer Meera Sodha, who was brought up on a range of delicious tidbits, from potato chaat to samosas: pastries stuffed with a range of fillings, from lightly spiced vegetables to curried chicken or lamb. Sodha calls samosas "triangles of joy."

In 2016, in a gap between an early breakfast and a late lunch, I visited a stall selling snacks at the organic farmer's market in Mumbai, run by a woman who had set it up after her husband died a couple of years before. That day, the stall was selling red *idli*, a kind of steamed savory pancake made from nourishing red rice. These *idli* were about the size of an American breakfast pancake but with a pleasingly rough texture from the grain. To go with it, the stallholder ladled out the most delicious chutney I'd ever tasted, made from coconut, lemon juice, salt, and green herbs, which was tangy and rich and sweet all at the same time.

For many in India, the greatest nutritional problem is not overnourishment but undernourishment. A few tasty snacks—which tend to be made from legumes such as chickpeas or grains such as millet—can help add much-needed nutrients, especially for the country's many vegetarians. If all snacks were like these tasty Indian morsels, it's hard to see much wrong with snacking. Like tapas in Spain or dim sum in China, traditional Indian snacks can form a pleasurable and sociable pattern of eating that potentially contains a greater variety of flavors and textures—as well as nutrition—than the old Anglo-Saxon model of three square meals.

While I was in Mumbai, however, I also started to hear about a new kind of snacks arriving in India. Weaving through the city's hot and crowded streets sitting in a black-and-yellow auto rickshaw, I noticed that alongside the old street stalls selling fresh oranges and coconut, there were newer kiosks crammed with cans of sugary fizzy drinks and sweets and rows of chips. In 2014, India spent $1.7 billion on savory snacks alone, and a 2015 market report announced that, although India was not buying as many potato snacks as other countries in Asia such as Japan and China, it was experiencing "exceptional growth" in this area. Likewise, chocolate bars, which were never traditionally eaten in India, are also now part of the landscape.[27]

The key to the rapid growth in India's chocolate market, according to insiders, has been "penetration and affordability." "Penetration" meant that when selling to India, multinationals focused not so much on selling to big supermarkets as on getting products into every tiny village store. The affordability part is that the multinational chocolate manufacturers created special small packages for India costing just a few rupees, cheap enough for a child to buy.[28]

I spoke to a woman who had spent time as an aid worker with children in rural villages near Bangalore. She told me that she had witnessed how enthusiastically commercial snack foods were welcomed by poor rural families. The families she saw often subsisted on little except white rice with small quantities of very watery vegetable curry. Their diets were low on protein—low on everything, in fact. When stores arrived in the village selling tiny packages of chocolate, the children were eager to try them. The value of these treats went beyond taste: they seemed like packages of Western affluence.

Unlike Ukraine, India is a country where people have always loved snacks and street food. But the arrival of these new commercial snack foods has been unprecedented, and their effect on public health has been a disaster. For millions of families, these packaged snacks have made the quality of a poor diet even worse. As with other aspects of the nutrition transition, this same pattern is repeating all over the world.

SNACK FOODS FOR THE WORLD

Barry Popkin can identify the year when snacking took off in China. It was 2004. Before that, the Chinese consumed very few snacks except for green tea and hot water. In 2004, Popkin suddenly noticed a marked transition from the old Chinese ways of two or three meals a day toward a new pattern of eating.[29]

The arrival of snacks in China has been a transformation of manners as much as of diet. A food culture based on sitting at a table to eat and waiting until you are hungry has morphed into one where people expect little top-ups of food and drink throughout the day.

In collaboration with a team of Chinese nutritionists, Popkin has been following the Chinese diet in snapshots of data every two or three years, conducting regular surveys of around ten thousand to twelve thousand people. Back in 1991, Popkin found that eating between meals was a rare activity in China. At certain fixed times of year, there were treats to supplement the daily diet. During the mid-autumn festival, for example, people would eat moon cakes made from lard-enriched pastry stuffed with sweetened bean paste. But such feasting foods were ritualized and rare, nothing like a casual cereal bar.

In 2004, out of nowhere, Chinese habits of snacking dramatically rose. The number of Chinese adults between nineteen and forty-four years of age describing themselves as eating snacks over a three-day period nearly doubled, while the number of children between two and six years old reported as eating snacks rose almost as much. By 2011, snacking had taken another massive stride forward. Popkin's data suggested that more than half the country was now regularly snacking. Based on the most recent data, more than two-thirds of

Chinese children now report snacking during the day. This is an eating revolution.[30]

The curious thing about snacking in China is that to start with, it actually made people healthier, because what people were snacking on, for the most part, was fruit. As of 2009, just 2–3 percent of children in China were snacking on sweets or other sugary foods, whereas 35–40 percent were snacking on fruit and around 20 percent were snacking on grains and nuts. This early phase of snacking was mainly a case of people—especially in the cities—having a bit of extra money to spend on food and being able to afford more of the foods they had always aspired to eat. The early data showed that Chinese children who snacked were actually *less* likely to be overweight than children who did not snack, perhaps because their diets had more variety to lighten the staple meals of rice.

Phase two of snacking in China has been very different. "The marketing comes in," Popkin told me, "and boom! boom! boom! the snacks are not healthy anymore." As of 2015, the commercial savory snack food market in China was worth more than $7 billion. Unlike in America or Europe, plain chips are not very popular, accounting for just 5 percent of the market. There is a huge appetite in China, however, for ultra-processed snacks of one kind or another, some salty, some sweet. The Japanese firm Calbee—the market leader for snacks in China, whose slogan is "Harvest the Power of Nature"—sells a range of deep-fried shrimp crackers and vegetable crackers and deep-fried extruded pea snacks.[31]

The rise in snacking is a significant part of a vast and dangerous transition in the way we eat, with a shift toward near-continuous grazing for many people. "There wasn't really any snacking in the world before the Second World War," Popkin tells me, at least "not on this large scale"—and not on this type of food.

Marketing has created new snacking habits all over Asia. Back in 1999, Thai people ate 2.2 pounds of commercial snacks per person per year on average, which only sounds like a small amount because at that time, Mexicans were eating 6.6 pounds per person and people in the United States an amazing twenty-two pounds. Frito-Lay—part of PepsiCo—saw an opportunity in Thailand. Frito-Lay knew what Thai shoppers did not yet know, which is that the Thai population were open to buying many more savory snacks, if only they could be marketed in the right way. From 1999 to 2003, Frito-Lay more than doubled the amount they spent on snack promotions in Thailand, with a range of TV campaigns tailored to different customers. Cheetos prawn crackers were pitched to young children, whereas Lay's potato chips were marketed to older, more affluent consumers. As for Doritos tortilla chips, a product that had no connection with traditional Thai food, the campaign was aimed at finding mostly new customers: to create an appetite where there had been none before.[32]

In 1999, Frito-Lay devoted ฿45 million to promote Doritos in Thailand. It gave out two million free samples and targeted teenagers with cool ads on MTV. The following year, the company quadrupled the marketing budget for Doritos and teamed up with Nokia to offer free mobile phones to consumers who could collect four jigsaw pieces from a pack of Doritos to complete a picture of a Nokia phone. Before long, salty chips became a normal part of the Thai diet.[33]

What's happening now with savory snacks in Asia happened in Western Europe in the 1980s. A Mintel market report on British savory snacks from 1985 noted the "explosive" growth of the snack industry, especially those aimed at children: items such as savory packets of Alien Spacers (multicolored maize produced by "direct extrusion") or Twirlers (extruded potato shapes in "entwined strands"). Chip commercials—many of which appeared during children's TV shows—were explicitly aimed at creating demand from the children themselves. The marketers knew that once the chips were in the house, the odds were that adults would also start to nibble them with alcoholic drinks and/or while watching TV, either before or after dinner.

What the snack marketers didn't predict was how many people would start to eat snack foods not alongside dinner but instead of it.

Because they offer a large amount of calories for not much money, commercial snack foods have helped to create a new face of hunger, one that coexists with obesity. In the old days, we knew that someone who couldn't afford a decent hot dinner was hungry. But now, much food poverty is disguised by the abundance of nutrient-poor snack foods, which enable people to eat thousands of calories without ever sitting down to a cooked meal.

"HE'S NOT REALLY BIG ON FOOD"

Snacking is not always about fun extras that you nibble before dinner. Sometimes there is a harsh economic logic behind it. For some hard-up eaters, packaged snack foods have become the thing you buy when you can't afford real food. In 2011, researchers conducted a series of focus groups about snacks with thirty-three low-income mothers in Philadelphia, most of them African American. Half of the mothers were people with obesity, and a fifth of them lived in food-insecure households that could not rely on having enough money to buy adequate nourishing fresh food. All of these parents fed their preschool children snacks of one kind or another. They gave their kids Danimals and Gogurts, Tastykakes and fruit cups, chips and cookies, and little boxes of candies like Mike and Ikes. These mothers were not thoughtless in their deployment of snacks. They spoke of trying to keep a handle on portion size by decanting value-size bags of potato chips into smaller sandwich bags, trying to ration the number of packaged cakes, and sometimes offering fruit instead, when they could afford it. But snack foods played a role in their difficult lives that ordinary food simply could not. The value of snack foods to these families was both economic and emotional.[34]

Snack foods are part of the answer to the question of how deprivation and obesity coexist. For those on low incomes, snacking is often a strategy for dealing with the need to skip meals to save money. A bag of potato chips is much cheaper than a plate of hot food in a café. Those who are food insecure—like the Philadelphia mothers—are more likely to snack than those who are not food insecure. Most snack foods eaten by American children are "energy rich and nutrient poor": high in sugar and other refined carbs but low in vitamins. Based on data from 2009–2010, children in the United States get 37 percent of their energy from snacks, but these snacks provide only 15–30 percent of the vital micronutrients their bodies need.[35]

On average, snack foods are far less nutritious than a regular plate of home-cooked food. And yet—unlike poverty foods of the past such as bread and dripping—snack foods inspire not anger or disappointment in those who eat them but a kind of gratitude and brand loyalty. They come in colorful packages, and each bite is infused with deep artificial flavors that satisfy on the tongue, if not so much in the belly. Those Philadelphia mothers saw snacks as a thing apart from food. To them, snacks were cheaper, simpler, easier, and more enjoyable than meals. One mother commented that her son would happily eat snacks all day, "but he's not really big on food."

Snacks have always been defined in contrast to meals, but it's an oddity of modern snacks that they are starting to be defined in contrast to *food* itself. A snack is everything that a meal is not. Meals are substantial, probably hot, savory, and ideally eaten in company. A snack is now generally something cold, insubstantial, eaten alone, and more likely than not to be sweet. No wonder many of us have started seeing snacks and food itself as two more-or-less separate things.

In their hard-pressed lives, the Philadelphia parents used snacks less as a form of nutrition than as a tool to manage a child's emotional state and sometimes their own. When talking of snacks, several of the mothers used the word *holdover*. A snack could be a "holdover" to stop a child whining too much between meals. It could also be a way to improve a child's mood and help a mother out in difficult situations such as a visit to the doctor. Snacks were things that appeased a grumpy child or rewarded a well-behaved one. One of the mothers commented that "if we have to go to . . . the public assistance office, we're in there for four hours, I better have candy, I better have some chips and some juice 'cause that's the only thing that's gonna manage him for three hours, or two hours."

One of the mothers said that she struggled to get her son to sit and eat a meal, whereas "I can get him to eat his snack in a second." Several of the mothers kept their snacks in a locked cabinet to stop both the kids and themselves from eating them all up. Meals, the mothers recognized, had to include vegetables and maybe some kind of starch or rice. Meals needed to be heated up, at the very least, even if they were not cooked from scratch. They required organization, labor, time, and expense. A snack, by contrast, could be eaten right "out of the wrapper." A meal was dreary and dutiful, whereas a snack was a joyous kind of reward that let everyone in the family off the hook for a moment. One of the mothers summarized the situation like this: "To me, the difference is that a snack would be something that my son wants and a meal, it doesn't matter, you want it, you don't want it, doesn't matter, you're gonna eat it."

It is only when you see how far the snack has risen that you can appreciate how low the meal has fallen in our collective estimation. For the Philadelphia families in the 2011 study, the main difference between snacks and meals was that snacks were loved so much more—whether by the mothers or the kids.

The rise of the snack food industry has been inexorable in part because the manufacturers have found a way to tap into many different markets at once. On the one hand, there are the poor consumers who use snacks as a cheap way of appeasing hunger. On the other hand, there are the middle-class consumers in China and Thailand—and elsewhere—who are using new pockets of spare cash to buy snacks to supplement an already rich diet. Both types of consumers find something in the snack that they could not find in traditional meals.

The main worry of snack food manufacturers, judging from market reports, is that we consumers will suddenly notice how bad for us all these snacks are and stop eating them. "There is a strong threat from substitutes—such as fruit," warns a Datamonitor report on savory snacks in the United States from 2015. As far ago as 1985, a Mintel report on British snack foods warned that sales of snacks might slow down because of consumer health worries, given that 52 percent of British adults claimed to be cutting down on fat and sugar. But this report also forecast that creating healthier alternative snacks might be a way for the industry to ward off the antisnacking mood. Whoever wrote the report can hardly have imagined just how right they were. So far from health concerns stopping us from snacking, it turned out that consumer diet worries could generate whole new markets for the snacking industry.[36]

THE HEALTHY SNACK

To step into the premium snack aisle of a quality American supermarket these days is a baffling experience. Here are "guilt-free" chips made of kale or blue corn, dried edamame nibbles, gourmet popcorn, and seaweed snacks in many varieties. You see fruit snacks that resemble chips, and chips that resemble fruit snacks. For those on gluten-free diets, there is jerky made from either beef or coconut and "super potassium snacks" whose true purpose is unfathomable. Even cookies now may boast of containing "ancient grains." These snacks claim to offer benefits that mere food cannot replicate and are priced accordingly. It's easy to spend more money on a couple of healthy snacks than you would on a hearty sandwich or a bowl of soup.

Each new health scare that emerges is both a threat and an opportunity for the snack manufacturers. The latest trend in nutrition is the "war on sugar." Data from Euromonitor International suggests that as of 2016 half of all consumers in a worldwide survey were actively looking for foods with no added sugars. As a result, healthy snack manufacturers have been moving away from conventional sugars and sweeteners such as high-fructose corn syrup. For many of us, a sugar-free snack that actually tastes good has become the Holy Grail. When our bodies experience some niggle or another, it's easy to assume that the answer is a better snack rather than no snack at all.

Much of what is sold as a healthy snack is, let's be frank, far from healthy. The sugar levels in "all-natural" snacks such as granola bars are often even higher than you'd find in a bar of chocolate. As of 2016, you could buy "yoghurt-coated strawberry fruit bites" from Tesco supermarkets in the United Kingdom marketed as a wholesome snack for children. Their sugar content was 70 grams per 100 grams, more than a Mars bar (at 60 grams of sugar per 100 grams). If you want a snack with more sugar per 100 grams than Ben & Jerry's chocolate fudge ice cream (27 grams of sugar per 100 grams), why not try pumpkin and chia seed "power balls" enhanced with gluten-free oats? I found a pack of these power balls recently containing 37.8 grams of sugar per 100 grams.

Healthy snacks only intensify the snacking habits of the world because we have lost our reason to stop nibbling. A generation ago in the West, people enjoyed snack foods but with a mild sense that you probably shouldn't eat too many of them because they were more unwholesome and greasier than a meal. The healthy snack takes our qualms away. "Go ahead and eat!" these products seem to say. "You'd be foolish not to." At the gym where I go, I sometimes look at the array of healthy snacks for sale and marvel at the chutzpah of the claims and the inflated prices. There are protein-boosted pancakes and ludicrously expensive juices that are supposedly energy-enhancing. But I can't help feeling that none of these items would be as restorative as a simple bowl of soup, plus the time in which to enjoy it.

There are signs now that our whole food culture is undergoing "snackification." As one consumer report notes, with the growth of more nutritious snacks, "dinner becomes a mere pause between other activities. Lunch is often scheduled out to accommodate an overflow of meetings and must-do's. And breakfast can be multitasked between commuting and working. . . . Snacks are no longer just whimsical, throwaway or anomalous moments; they are an essential part of how we eat."[37] Snacks are both cause and consequence of our rapidly changing food culture.

Perhaps we ourselves, as much as the food we eat, are becoming "snackified." We act as if we are eating and drinking on the fly even when we are not. Many of us have been encouraged by snack-food manufacturers and retailers to nibble our way through the day as if we were at a never-ending theme park. Have you noticed that often in a coffee shop now, someone will get to the front of the queue and ask for their latte in a takeaway cup, only to sit down and drink it in the café? It's as if a ceramic cup and saucer feels like too much of a commitment.

Here's a thought. Imagine if we took those calories from snacks and spent them instead on food for dinner. We could eat rich, hearty, satisfying meals like Olia Hercules (if only we could find the time to cook them). Or another option would be to take the calories from snacks and not eat them at all. Remove all the snack foods (including the so-called healthy ones), and maybe—just maybe—the levels of obesity and diet-related ill health might start to be reversed. The question is, what would we do with ourselves and our days if our snacks were gone? It is startling to think that snacking may form a bigger element of what the average person eats in a day than any single meal.

I don't happen to believe that eating between meals is a moral disaster. As someone who sometimes used to eat chips for breakfast as a teenager, relishing every salty bite, I am in no position to judge. Some people may flourish best on six or eight small meals a day rather than three big ones. But the whole economy of commercial snacking has spiraled out of control. What is missing, in the cacophony of snacks, is the experience of not eating, which used to give meals their purpose. Without silence, there is no music. If we never stop eating, there can be no true meals.

We are moving further and further away from the old eating culture of big, filling, sociable meals punctuated by breaks. Snack foods have made our day of eating extend almost indefinitely, disrupting the old rhythms of life. Breakfast, lunch, and dinner used to give our lives a focal point. They marked the start and end of the day. Meals were how we came together and how we celebrated. They imposed certain rules on us about what to eat and how to behave. The never-ending snack comes with no such structure or rules. It can be eaten wherever, whenever, with no other human being judging whether what you eat is right or wrong.[38]

EATING ALONE TOGETHER

From 1970 to 2012, the number of American men living alone doubled, from 6 percent to 12 percent, according to data from the Census Bureau.[39] As the *New York Times* reported in February 2016, 62 percent of American professionals say they eat lunch at their desk every day, and some eat every meal there. For many millions of busy solo diners around the world, the experience of eating is completely at odds with the sociable images of meals that the media bombard us with.

Hence the phenomenon from South Korea of people watching internet videos of beautiful people eating—a remote form of companionship. This is called *mukbang*, from the Korean words for eating (*meokneun*) and broadcast (*bangsong*). The stars of these films—BJs, short for broadcasting jockeys—may earn as much as $10,000 a month from eating food in front of a webcam while chattering about the experience in a friendly way and making loud slurping and nibbling noises. Most South Korean mukbang are watched through Afreeca TV, an online channel. The money comes from viewers sending their favorite BJs "star balloons," a form of virtual currency that can be converted back into actual money.[40]

A typical mukbang video will show a slender woman eating unfeasible quantities of fast food. More than 2.5 million people have watched Yuka Kinoshita, a South Korean BJ who looks like a tiny Manga cartoon princess, eating two packs of instant packet katsu curry with ten and a half pounds of steamed rice, amounting to 6,404 calories, enough energy for an average woman for more than three days in a single sitting. "I forgot how yummy rice and curry is," she comments at one point. "An unbeatable combination!" Kinoshita's other videos include her eating ten packs of Kraft Macaroni and Cheese, six and a half pounds of Oreo cereal, six packs of instant ramen, and more than a hundred pieces of sushi.

It's hard to pin down what makes these videos quite so compelling. Once, the prospect of watching someone else eat and make banal comments for twenty minutes would have been considered as exciting as watching paint dry. But mukbang clearly strikes some kind of voyeuristic chord, judging from the huge audiences it generates. Trisha Paytas, an American mukbang "star," gets more than a million views for her YouTube videos featuring herself eating vast quantities of Shake Shack burgers or fried chicken from KFC. Mukbang seems to liberate people to confess their own anxieties and desires about food. In a world where most of us eat too much, these videos of gargantuan feasts can make us feel that our own dinner of takeaway pizza is modest.

Food is so omnipresent in our real lives that you might think we would want a break from it online. Yet after a day of snacking and grazing, we seem to have an unquenchable hunger for digital food that hasn't been met by our actual meals. It isn't just mukbang. Speeded-up recipes featuring hands but no faces have become some of the most popular items on the whole internet. The appeal of these cooking videos stretches far beyond the small subsection of the population who might buy cookbooks or read a recipe column in a newspaper. In a world of celebrity and bad news, it seems that what many of us crave is the soothing sight of unknown hands doing things to food. Almost all of the most viewed cooking videos are being generated by Tasty, an offshoot of BuzzFeed. In under a year, from 2015 to 2016, Tasty became BuzzFeed's most popular Facebook page.[41] In June 2017 alone, Tasty's videos were seen by 1.1 billion people.[42] Most likely you've shared one of its cooking videos yourself without necessarily knowing where it came from: "8 Game-Changing Pizza Recipes," "9 Desserts for Peanut Butter Lovers." Many people seem to know them simply as "those food videos with the hands."

When you watch a Tasty video, you have a strange sense that it is your own hands doing the cooking: cracking the eggs, melting the butter, twisting the puff pastry into neat shapes, slicing cucumbers and avocados for a salad, squeezing lemons, ladling out a spoonful of hearty winter soup. Crucially, we never find out who the hands belong to. BuzzFeed has discovered that viewers don't like to see hands with too many distinguishing marks or jewelry. Viewers may be "triggered" by the sight of a bracelet because we don't want to be disabused of the impression that it's our own hands that are chopping those onions so speedily.[43] At the end of every Tasty video, you get a little dopamine hit from the sense of a task being completed in a clean and tidy kitchen: the final garnish of spring onions scattered on a dish of noodles, a white glaze zigzagged on top of a pastry, a scoop of ice cream on an apple pie. No loose ends have been left untied.[44]

The success of both Tasty and mukbang shows how socially and temporally disconnected many of us have become in our meals. "Let's eat together next time" is a common way to say goodbye in South Korea, but much of the time we are not eating together. In South Korea, as in many other Asian countries, there are strong social taboos against solitary eating, yet given the rising numbers of unmarried people, solitary meals are a reality for millions. The "social eating" of mukbang can make someone feel less self-conscious about eating bibimbap for one. One of the executives at Afreeca TV, Ahn Joon-Soo, has said that "even if it is online, when someone talks while eating, the same words feel much more intimate."

I spent too many of my teenage years alone at the kitchen table, eating to fill a void, feeling terrified and ashamed. I sometimes wonder how different my experiences of food would have been if smartphones had existed then. Would I have found kindred spirits online to make my secret appetites feel less shameful? Might I even have found enough companionship from my screen that I didn't feel the urge to binge so often?

It's not that eating alone has to be lonely. Mukbang could be a sign of a food culture that is slowly adjusting to the reality that large numbers of people do not eat in traditional family units. Given the rise of solo living, meals eaten alone are simply a fact of life for millions. According to Euromonitor, the number of people living alone worldwide rose from 153 million in 1996 to 277 million in 2011. Around a third of all households in the United States and United Kingdom have one person living in them, yet cookbooks (except for diet books) still routinely give recipes for four to six people. I spoke to a single woman in her fifties who said she loved eating out alone, treating herself to a lovely restaurant meal. But her married friends would often make negative comments such as "Why bother, when it's only you?" which upset her.

If only we celebrated them more, meals for one could be liberating, says Signe Johansen, the Norwegian author of the 2018 cookbook *Solo: The Joy of Cooking for One*. Johansen notes that when cooking for ourselves, we can feel freed up from the "exhausting performance" of trying to impress others with food. Johansen loves the fact that she can "add as much garlic and chilli to a dish as I damn well please." Like Edouard de Pomiane, Johansen relishes short bursts of cooking time as a way to unwind at the end of a working day. She pours herself a whisky and soda, switches on a podcast, and feels happy at the "prospect of rustling up a simple meal."[45]

We continue to idealize the experience of eating in groups, but even people who don't live in one-person households are often trapped in new feelings of loneliness. As Sherry Turkle documented in her book *Alone Together* (2013), many people feel greater intimacy toward the iPad in their hand than the people in the room with them. Social media also encourages us to interrupt our own meals to photograph them.

It's one of those rituals that seemed strange and rude only a couple of years ago and now is becoming almost normal, at least among millennials. The meal starts, but no one takes a bite. There is a collective intake of breath, as if to say grace. Instead, everyone takes their cameras out and leaps up to photograph the food, ready to post it on their Instagram feeds. Hashtag food. Hashtag friends. Hashtag fun. There are those who now record the latte art on each cup of coffee with the same loving care that was once displayed only by parents taking photos of a baby's first steps.

The process of sitting at a table over a plate of food, making eye contact and conversation with other human beings, is now completely out of sync with the way most of us spend the rest of our days, responding to the tyrannical bleeps from our phones. The more time we spend with our virtual companions and virtual meals, the less energy we may have in reserve for our physical meals eaten with flesh-and-blood companions.

It's impossible to gauge the full impact of smartphones on our eating lives, because academic studies are slow, whereas the changes wrought by novelties such as BuzzFeed's Tasty videos are happening in a span of months. But the early signs are that screens and meals combine about as smoothly as oil and vinegar—that is, not at all. One study from 2014 looking at American teenagers found that those who used screens at family mealtimes were less likely to eat green vegetables or fruit, more likely to drink sugary sodas, less likely to talk to family members, and indeed less likely to see family meals as important at all.[46]

The internet was always a strange place to go in search of home. The cozy food we get from our screens can hide from us the extent to which we have moved away from traditional meals and all the benefits they gave us. No wonder we are so in thrall to those clever hands on Facebook whipping up meals in a trice. These videos allow us to imagine that we, too, are floury-handed craftspeople in the kitchen rather than busy people whose thumbs are now used almost exclusively for swiping and tapping a small nonnutritious object.

Many of us are tantalized by the idea of home-cooked food nowadays, surrounded by delectable things yet feeling that we have no time to cook them or properly enjoy them. This is sad. The great irony of our collective belief that we lack the time for proper meals is that nothing makes you feel so rich in time as a good meal, especially if it is shared. When we obsess too much about time efficiency, we enjoy our time less. Research on how we experience time suggests that we actually feel less harried when we stop holding on to our minutes and seconds so tightly and start freely giving more of our time away—by cooking dinner for someone we care about, for example.[47]

Even in the rushing and bleeping of modern life, there are still moments when time seems to become elastic and expansive. Most of these moments, in my experience, are spent sharing food. It's a summer evening; the cherries you bought are large and lush, and you eat them lazily, until your mouth turns inky-red. There's a pot of fresh mint tea on the table, and you share out the final slice of almond cake between everyone. It is as if someone has handed you a sliver of time, an excuse to stop counting the minutes for once and actually experience them.

CHAPTER FIVE

THE CHANGEABLE EATER

HOW I CAME TO EAT SKYR FOR BREAKFAST I CANNOT say. For most of my life, I was unaware such a substance even existed. But something must have convinced me that skyr was a good thing, for here I am, one summer's day, spooning it up with strawberries and toasted hazelnuts as if it were a perfectly normal thing for me to eat. Which it now is.

For the uninitiated, skyr is a fermented dairy product from Iceland, somewhere between yogurt and cream cheese in texture. It feels as rich on the tongue as mascarpone yet is remarkably low in fat and high in protein compared to other cultured dairy products. Skyr has been eaten in Iceland in some form or other since Viking times. More solid than yogurt, it will not fall off a spoon if you twirl it (unless you do what my youngest child did and tap extra hard on the spoon while aiming it at the floor, to prove a point). It is pronounced "skee-er" not "sky-er," as I first thought. Ten years ago, skyr was barely spoken of outside Iceland, but as of 2016, the global market for skyr was worth $8 billion and growing, an astonishing transformation.

So many new edible wonders have arrived on our plates in recent years that we almost forget to be surprised that they are there. I find myself cooking quite casually with sumac and dried ground Persian limes, Middle Eastern seasonings that until very recently were completely unknown to me. One day, decades from now, my grandchildren will ask me how old I am, and I will reply, "Old enough to remember a time when I didn't know how to pronounce quinoa."

Each year, culinary forecasters come up with a relentless new list of arcane food crazes that are supposedly poised to happen any moment now. Chai-flavored cookies! Blue-green algae! Spicy nduja sausage! And sometimes the latest trend isn't a new food at all but something very old. Kale, for example. In 2009, chef Dan Barber published a recipe for Tuscan kale chips crisped up in the oven using olive oil that changed the way that many people thought about this cabbagey vegetable. Around the same time that Barber was experimenting with overcooking kale, another American chef, Joshua McFadden, was playing around with not cooking it at all, but shredding it raw with garlic, oil, and chili flakes. McFadden's raw kale salad became one of the most copied recipes of modern times. By 2017, kale chips were in every supermarket, kale salad was in McDonald's, and raw kale was generating more than $100 million a year in US grocery sales.[1]

Over the past five years, eaters have become far more changeable in our behavior around food. In the spring of 2017, I sat eating bright golden turmeric hummus with Susi Richards, who at that point was working as head of product development at Sainsbury's, the second-largest British supermarket chain. "Turmeric is a trend that came out of nowhere," she remarked. Sainsbury's was already selling fresh turmeric and powdered turmeric, turmeric tea, and little bottles of "turmeric shots" to appeal to people who had heard about the anti-inflammatory health benefits of curcumin, the main compound in turmeric. Now, Richards was taking a punt on turmeric in hummus, although she conceded that the neon-yellow color was not for everyone.

During the twelve years she was at Sainsbury's, Richards noticed big changes in the way people choose food. "Customers are acting much more erratically than they were before," she told me. On the one hand, lots of people want vegan health foods boosted with sweet potato and avocado. On the other hand, she witnessed a rise in "dude food"—swaggeringly meaty convenience foods such as nacho pizza, designed to be eaten on a Friday night in with beers. Around 2014, Richards noticed that shoppers were "much more demanding and worldly wise" than they used to be. It was becoming harder for supermarkets to tell who was a traditionalist and who was health-conscious. In just the past three or four years, it has all become more of a mash-up. Data from the Sainsbury's Nectar reward cards (which track what customers are buying) reveals that many of them assemble baskets of food that seem to belong to multiple different diets. A person may equally fervently want both a superfood beetroot salad and a sticky toffee pudding. If food is identity, then many of us now have split personalities.

This extreme trendiness—or faddiness, depending on your point of view—has been the latest stage in a longer process of change in how we obtain our food. In the distant past, our eating habits were limited by the ingredients that could be grown on the land nearest to us. Then, for generations, children sat captive at the table and either ate or rejected whatever their mothers wanted us to eat. Now we can eat whatever we have money to buy: a freedom that is both heady and unsettling.

Food trends can seem the most trivial and pointless of subjects. When I hear someone announcing that black charcoal smoothies are about to take the world by storm, I roll my eyes and continue to sip my comforting mug of hot tea, which I have been making in much the same way for decades, and which I hope to carry on drinking, preferably out of the same mug, until the day I die. When Hawaiian sashimi poke bowls are advertised as "the new sushi," I think, "What was wrong with the old sushi?"

Yet, however frivolous they may seem, food trends have the power to affect the way we all live, even when the trend is for something as seemingly inconsequential as avocado toast. Our changing tastes have consequences, both for us as consumers and for the people who produce our food. Very often, the cult of fashionable foods is not good for either farmers or eaters. It's safe to say that anything advertised as a superfood is not as uniquely beneficial as it pretends to be. Unexpected things happen when millions of eaters shift our diets all at once. It's like when too many people stand up at one end of a rowboat and the other end capsizes.

What we are talking about here is a series of fundamental changes in human behavior, as much as in ingredients. Through trade and empire, humans have welcomed new items into their diets many times before, but this process of exchange used to be slow and incremental. We were largely satisfied with food that stayed the same from year to year, as long as there was enough of it and it tasted good. Today, as never before, populations have become erratic in their choice of foods, expecting to change what we eat for dinner almost as often as we change our socks. Perhaps the biggest change is our addiction to change itself.

WHAT'S SALSA?

Why do we now eat so many new foods, in so many different forms? In his 2006 book *The United States of Arugula*, journalist David Kamp marveled at the speed with which novel foods were adopted into our diets. Kamp recalls someone telling him at a wedding in 1984 that he needed to learn about salsa, to which he replied, "What's salsa?" By 2006, salsa had overtaken ketchup as the most popular condiment in America.[2]

Yet now, in terms of food trends, 2006 seems like ancient history, because this was life before Instagram, the photo-sharing service created in 2010, which has hugely accelerated the passage of food trends around the world. In those innocent days when Kamp was writing his book, people were still selecting new foods based on how they imagined they might taste, rather than heading to the café with the most "instagrammable" brunch. Food trends have never moved so fast or spread so extensively around the world as they do today, and part of that is to do with the influence of social media. It took decades for extra-virgin olive oil, pesto, and the arugula of Kamp's title (rocket in Britain) to stop seeming like elitist foodstuffs and find their way into the shopping baskets of millions. By contrast, with Instagram and other social media sites, a chef's bright idea can spread in months or days.

The speed at which recipes and new ingredients travel is faster now than even five years ago, according to Nidal Barake, who runs a food innovation company in San Francisco called Gluttonomy. Barake tells me that he observes certain recipes—from pulled pork to za'atar chicken—"snowballing" from restaurants to casual coffee chains to home kitchens.

So great is the power of social media to influence what we eat that many cafés and restaurants have started designing their menus and lighting and tableware around customers' desire to photograph and "share" the meal online. Food arranged in photogenic bowls (otherwise known as bowl food), salads decorated with random scatterings of flowers (which may or may not be edible), unusual colors of vegetables—these are just some of the trends that have been driven by social media in recent years.

As well as promoting new foods—from acai to hemp milk—social media has also succeeded in resurrecting the popularity of some ancient ones. Take eggs, which for decades were shunned by health-conscious people because of the high cholesterol in the yolks but which are now trending on social media. From the 1990s onward, a series of studies suggested that eggs were not to blame for heart disease, after all. The cholesterol in yolks does not—as previously thought—convert straight into blood cholesterol in the body. This was great news for those of us who have always regarded the egg-white omelette as a sad nonsense. Yet sales of eggs remained very low in much of the world (with the exception of Spain, where the tortilla remains god). It has only been with the endless brunch photos on Instagram that egg sales have finally started to rise again.

As of 2016, #eggs was the eighth-most-hashtagged food on Instagram worldwide, behind #pizza, #sushi, #chicken, #salad, #pasta, #bacon, and #burger, but more than twice as frequently hashtagged as #sandwich, #noodles, or #curry.[3] As an Instagram food, eggs have many things going for them. Eggs are compatible with the high-protein and low-carb diets popular on social media, and they are #vegetarian if not #vegan. But maybe most importantly, the color contrast between white and yolk makes eggs very photogenic. In 2017, one of the leading Instagram food trends was the "cloud egg," made from soufflé-whipped whites baked with yolks inside, to look like a cartoon version of a fried egg.[4] In the iconography of modern food photos, eggs have become something like Cézanne's apples or Matisse's oranges: rounded symbols of happiness.

In some ways, all this "shared" food has a certain magnificence. As of 2017, there were more than 250 million posts on Instagram with the hashtag #food. You can go on Instagram and see how people live and eat in different cities and distant continents. It is possible to compare a breakfast in Helsinki with one in Nepal or to see the way that oranges are green in Vietnam but orange in London. It is heartening to be reminded how many people are eating, and being soothed by, delicious variants of soup at any given time or how many waffles are being made in how many millions of kitchens. Food has always been a way to make connections, and through online food, we can gather thousands of glimpses of other people's lives and thousands of new ideas about food.

Thanks to all this online sharing of ideas, ordinary home cooking has become far more experimental and open-minded than it ever used to be. I sometimes think I have learned as many new cooking techniques from blogs and social media as from my whole library of cookbooks. I have learned that bread dough does not always have to be kneaded and that scrambled eggs can be cooked in water instead of oil (an idea that came from the Californian chef Daniel Patterson). I have learned how to make Indian butter chicken with minimal effort and time in a pressure cooker. Best of all, I learned from J. Kenji Lopez-Alt of the Serious Eats website that asparagus—my favorite vegetable—is most delicious of all not steamed, as I'd always believed, but braised, to bring out the sweetness. But if blogs and social media have been great vehicles for spreading useful new ideas about food, they have also been a medium for spreading bad ideas.

By its very nature, Instagram demands that we focus on the appearance of food at the expense of its taste. It prioritizes dishes that are colorful but bland (smoothie bowls topped with flowers) over those that are shapeless and nondescript in color but delicious and nourishing (a stew, for example).

At its worst, social media promotes foods that are toxic or unwholesome, simply because they look pretty. A generation ago, it was accepted wisdom that consuming too many synthetic food colorings was a bad idea because some of these chemicals could trigger allergies and, in any case, added no nutritional value. But now, food colorings are back with a vengeance with the rainbow foods of Instagram, notably rainbow bagels. What's a rainbow bagel? It is, as one online critic puts it, "a dense, overworked, phosphorescent ring of disappointment."[5] A rainbow bagel is made from dough dyed with seven different food colorings and baked into a ring-shaped bread roll that would look too garish even at a five-year-old's birthday party. Only on Instagram could such a thing seem better than normal bread. It is like an idea of joy rather than actual joy.

Like our food culture in general, the food on social media is polarized between foods that are labeled unhealthy and those deemed to be healthy. Many of the trends on Instagram—the rainbow bagel being a case in point—have been for extreme junk foods such as ice cream sundaes garlanded with cookies and marshmallows or messy hamburgers with so many layers of meat and cheese that they are actually impossible to fit in a human mouth. But at the opposite end of the scale, Instagram has also made a fetish out of healthy foods, especially for artfully arranged and virtuous breakfasts involving berries and bowls of either overnight oats or chia pudding or some kind of yogurt.

Enter #skyr, which at the time of writing has more than a quarter of a million mentions on Instagram as a food that fits perfectly with the vogue for broadcasting the fact that you are eating an aspirational and vaguely Scandinavian breakfast. In just a handful of years, skyr went from an unknown food to the mass market, and it was helped by social media, which spread the message that this Icelandic fermented dairy food was higher in protein than yogurt. In 2006, Siggi Hilmarsson, an Icelandic entrepreneur, was selling his Siggi's skyr in just two outlets in the United States. By 2016, Siggi's was in twenty-five thousand outlets. That was the year that Siggi's vanilla-flavored skyr made it into Starbucks cafés and became a normal part of breakfast for millions of Americans.[6]

In Britain, skyr became established even faster. It went from obscurity to mainstream in the space of just two years, from 2015 to 2017. Before June 2015, skyr had never been sold on a commercial scale in Britain. That month, two separate firms launched their versions of skyr in the United Kingdom. One of these, the more delicious one as it happens—Hesper Farm Skyr—was launched in Yorkshire by a twenty-one-year-old farmer named Sam Moorhouse, who had never seen, tasted, or even heard of skyr until 2014. But like many millennials, Moorhouse is open to new tastes. He tells me he happened on skyr as a way to extract better profit margins from his family's herd of Holstein cows than by selling plain milk. Meanwhile, the European dairy giant Arla Foods (who also makes Lurpak butter) was having the same thought. Arla's mass market skyr can now be found in all the major British supermarkets, where it sits next to the Greek yogurt as nonchalantly as if it had always been there.[7]

One of the great mysteries of the food business is how this process of trends actually works. Even in this age of novelty, not every new food is so readily accepted by consumers. Like skyr, some new items find instant success, while others fail to take off.

NEW BUT NOT REALLY

Nothing looks as ridiculous or strange as a food trend that hasn't quite happened. For every new food that is welcomed onto our plates—from kale chips to ras el hanout spice mix—there are thousands more almost-trends that never fully gain traction. The cappuccino-flavored chip. Purple ketchup. Blue cheese–flavored Portuguese custard tarts. (Strangely, I did not make these up.) In 2014, more than fourteen thousand new food and drink products were launched in the United States, but the vast majority of these products will soon disappear without trace.[8]

In the winter of 2016, I strolled around a food trade exhibition in east London featuring new products from hundreds of exhibitors, most of them pushing items that the consumers of the world—including me—had not yet discovered a taste for. Here was avocado ice cream and chewy candies made from baobab fruit, a superfood. There were so many iterations of "the new coconut water" that I lost count. One man was promoting maple water. "We are like coconut water with half the sugar. People love us, they can't get enough!" he added, cheerily, perhaps hoping that if he said it often enough, it would come true.

Food trends are notoriously hard to call because what we are talking about is the vagaries of human desire, which are easy to influence but hard to manufacture. You can't create a food trend just by announcing it. In the years before it is fully accepted, a new food often looks outlandish and even revolting.

It is perfectly natural for us as humans to be suspicious of, if not outright disgusted by, new foods. According to some psychologists, this disgust would have originally had an evolutionary benefit, to keep us safe from ingesting toxins and dangerously rotting foods when we were hunter-gatherers. When it comes to eating, humans have always been poised somewhere between neophobia, fear of the new, and neo-philia, love of the new. We don't want to eat a poisonous berry and die, but neither do we want to miss out on all the good stuff that can give our bodies energy and our tongues pleasure.[9]

The truly successful food trends are those that offer something new to excite our neophiliac side while also appeasing our cautious neophobic side with a sense of the familiar. This partly explains why skyr took off so rapidly. When it arrived on the supermarket shelves, we were ready and waiting. Skyr is really just Greek yogurt dressed up in a Nordic woolly jumper.

"It's new but not really," is Lynn Dornblaser's take on the skyr trend. Dornblaser is the director of innovation and insight for Mintel, a global market research firm, where she has worked since 1986. She speaks with the cheerful world-weariness of one who has seen food fads come and go. "A trend can't be too new," she tells me, because it has to make sense to people. What Dornblaser has learned over the years is that very few "new foods" are in fact truly novel. "The thing I find so funny about skyr is that it's just yogurt."

Many of the most successful food trends are not pure innova-tions but piggyback on a flavor or ingredient we already have a liking for. The sudden success of skyr would have been unimaginable with-out a preexisting trend for Greek yogurt. Market insiders see skyr as part of a war of "origin yogurts" from different countries.[10] Bulgarian yogurt is huge in Singapore and Thailand, often in drinkable form, and creamy Australian-style yogurt is becoming popular in North America. And now we have Icelandic skyr. But none of the origin yogurts can come close to Greek.[11]

American yogurt first went fully mass market in the 1970s with Dannon, whose advertising alluded to the longevity of yogurt-eating centenarians in Georgia. "Yogurt then tended to be whole milk and didn't have many thickeners," Dornblaser remembers. It felt like something vaguely macrobiotic and tied in to the American counterculture.

Then, in the 1980s, yogurts became much more dessert-like, laced with thickeners, emulsifiers, sweeteners, and flavorings from plain old strawberry to such exotica as cheesecake flavor or chocolate chip. "Yoplait drove that market," says Dornblaser. Yogurts (and frozen yogurts) became the dessert it was still acceptable to eat on a diet. In all developed countries, from the 1980s to the 1990s, packaged yogurt stopped being a minority taste. Yogurt was trending.

At this point, an Indian or Polish or Turkish person might object that yogurt—and fermented dairy in general—is not a trend but a staple food. In many cultures, yogurt does not come in plastic tubs with fruit flavors but is made fresh at home most days in pans and jars. But in Western countries such as Britain or the United States, we had different dairy traditions and so the big food manufacturers could sell yogurt to us as a "new" food. They told us these sugary yogurts were healthy, and we believed them.

It was Chobani, the market leader for Greek yogurt in the United States, that changed the rules of the yogurt market around the world. In seven years, its sales grew by $1.8 billion. Despite costing around 25 percent more than regular, thinner yogurt, Chobani strained yogurt managed to gain around a third of the entire US yogurt market in a matter of a few years. Its success was due in large part to the entrepreneurial brilliance of Hamdi Ulukaya, who missed the yogurt he had grown up with in Turkey. As a twentysomething student in New York City, Ulukaya could not understand why Americans tolerated such bad yogurts: oversweetened and bulked out with various unpleasant thickeners and deadened with preservatives. From his Turkish childhood, he knew how much better yogurt could be. True yogurt, Ulukaya knew, should be live with bacteria, not zapped with preservatives that killed those bacteria. But he also realized that an American shopper who had never tasted good yogurt would not think that way. Not yet.[12]

In creating Chobani, Ulukaya did everything he could to make this "new" strained yogurt appear familiar to an American consumer.

Ulukaya felt that the existing leading brand of Greek yogurt—
Fage—had poor branding. Shoppers were confused about how the
name should be pronounced (the *g* is pronounced soft, like a *y*). More
fundamentally, they did not know what this Greek yogurt was for.
Was it a thick and tangy treat? Or was it a diet food? And why was it
so expensive? When selling Chobani, Ulukaya emphasized the high-
protein content to explain to undecided shoppers why they should pay
more. Given his own Turkish heritage, Ulukaya could have called it
Turkish yogurt, but he stuck with Greek yogurt instead, because this
was a more recognizable concept. Finally, he mixed the yogurt with a
series of comforting American flavors such as blueberry. As Rebecca
Mead wrote in *The New Yorker* in 2013: "With Chobani, Ulukaya
has transformed a product with a distinctively ethnic identity into an
entirely American product."[13]

By the time that Siggi's skyr started to appear in American grocery stores in 2004, much of the work of persuading consumers that
they would like this new product had already been done by Chobani
Greek yogurt. Dornblaser feels that with skyr and Greek yogurt, the
American relationship with yogurt has come full circle. In the 1960s,
when making your own yogurt was part of the health-food counterculture, yogurt was something sour and wholesome. Then it became
something sweet, heavily marketed, and ultra-processed. Now it is
becoming more sour and wholesome again.

When I asked Dornblaser to summarize food trends over the
past three decades, she said the biggest trend has been "the change
in our notions of what healthy is." Thirty years ago, "diet meals" in
the United States would be low in calories but might also be high
in sodium and cholesterol and low in fiber. "We didn't know any
better," she said. This started to change in 1989 when the CEO of
Healthy Choice foods had a heart attack and complained in a board-
room meeting that his company didn't sell "anything I can eat." "But
now it feels like the focus is a much more holistic approach to health.
Consumers want to eat the very *best* food. The emphasis on health

and wellness has changed everything about how we eat." Skyr is just one small element in this larger trend.

For the Icelanders who have made skyr such a central part of their lives for so long, it must be strange to see the rest of us all at once "discovering" it as a high-protein snack to eat with our morning coffee. But many food trends are like this. One person's trend is another person's long-held culinary tradition.

Another of the most trumpeted food trends of recent years has been for the food of Africa.[14] Numerous excitable articles have praised African food in Europe or the United States or Australia as a hot new fashion. This is all very well except for the small details that, first of all, there is no such thing as "African" food (as opposed to Ghanaian food or Tunisian food or a host of other regional cuisines, ways of eating that are as varied as the continent itself), and second, you can't really call food that is already eaten by 1.2 billion people a hot new thing.

The whole phenomenon of food trends is part of a bigger problem in our food system, which is the way that we experience ingredients as disconnected from their place of origin. When a food becomes suddenly trendy, this will inevitably have far-reaching consequences for the people who produce it, for good or ill. But we eaters are mostly blind to such consequences, because we have been encouraged to think of the foods we consume as things that appear on our plates, as if by magic, simply because we desire them.

NOW COMES QUINOA

During the whole of the 1950s, the *New York Times* mentioned quinoa just once. The year was 1954, and the headline was "Now Comes Quinoa." A tiny item on the news pages noted that the US Department of Agriculture was "tinkering with the thin leaves of the quinoa plant" as a possible spinach substitute for Americans. The anonymous author found that "the tender young shoots serve as salad greens." He or she did not recommend eating quinoa seeds, however, complaining that they tasted "something like soap."[15]

To say that tastes change is an understatement. Between 1990 and 2018, the *New York Times* ran more than three hundred stories in which quinoa featured, and they were all very definitely about the seeds and not the leaves. Cookery columnists shared their excitement about rediscovering this grain-like seed that was as wholesome as brown rice but higher in protein. And gluten free!

By the early 2000s, among healthy eaters, quinoa was close to becoming a staple. From 1961 to 2014, quinoa production in Peru increased from 22,500 tons to 114,300 tons. Quinoa salad is now a byword for healthy eating all over the world. I've seen it in Cape Town and in London, in Mumbai and in Brussels.

Back in the Andes, where the seed has been cultivated and eaten since ancient times, quinoa consumption is declining. In the southern altiplano of Bolivia, where quinoa is grown for export, many of the farmers are no longer eating their own quinoa because it simply costs too much. Sven-Erik Jacobsen is a professor of life sciences in Copenhagen who has been doing fieldwork on quinoa in Bolivia for more than twenty years. Jacobsen has observed firsthand how the world's runaway appetite for quinoa has affected the lives of those who used to eat it as a staple food in Bolivia. In the year 2000, the price of quinoa was US$28.40 per one hundred kilos. By 2008, it was $204.50, an increase of more than 600 percent.[16]

Because of this huge and sudden price hike, many Bolivians can't afford to gain the nutritional benefit of their own quinoa. During the same period in which world production of quinoa tripled, Bolivian consumption went down by more than a third. It's now cheaper and easier for Bolivians to eat instant wheat noodles than to buy the staple carbohydrate of their own land. Aid workers fear that the thirst from health-conscious North Americans for super-nutritious quinoa to add to their already healthy diets may lead to a rise in malnutrition in Bolivia, as locals increase the amount of poor-quality overrefined carbohydrates in their diets.

The rising demand for quinoa has left its mark on the land. The urgency of producing vastly more quinoa in the area of Bolivia around the Uyuni salt flat has completely changed the way that farming is done. It used to be that quinoa was farmed manually, using methods that were slow and laborious but sustainable. Now, increased use of tractors has led to soil degradation. There used to be llamas here too, whose manure supplied fertilizer for the soil, but now the llamas' pasture has been moved to make way for bigger quinoa fields. The whole relationship among farmers, animals, and land has shifted in this part of Bolivia, all because of our perfectly blameless-seeming desire for vegan salads of quinoa and roasted root vegetables.

Growing more quinoa is not a bad idea in itself. In 2013, the Food and Agriculture Organization at the UN declared it the International Year of Quinoa. Quinoa, the organization trumpeted, was "the only plant food that contains all the essential amino acids, trace elements and vitamins and contains no gluten." In theory, quinoa could be the food to feed the world: a solution to hunger that addresses the quality of what the poor eat rather than just the quantity.[17]

With the population of the world edging toward nine billion by 2040, quinoa looks like a sustainable and nourishing solution to the question of how to feed all these extra mouths. As a crop, it is tolerant toward both drought and cold. This may be vital, given the environmental fluctuations of climate change. It is also the only plant food that is a complete protein. Other vegan sources of protein, such as lentils and beans, need to be combined with a starch such as rice or bread to become nutritionally complete.

"If you ask for one crop that can save the world and address climate change, nutrition, all these things—the answer is quinoa," Sven-Erik Jacobsen has said. However, farmers in Bolivia have been very reluctant to share the country's gene banks of quinoa seeds with the world. As journalist Lisa Hamilton found in 2014 when she traveled to Bolivia to interview quinoa farmers, ownership of quinoa seed is for many Bolivians a question of food sovereignty. The great fear is that if Bolivia opens up its quinoa seed to the world, it will become marketed by US companies and no longer belong to Bolivians.[18]

Maybe quinoa will still be the food to save the world. Over the past couple of years, there have been forays into more locally grown quinoa, in the United States, Canada, and the United Kingdom, where Hodmedod quinoa is grown in the county of Suffolk.

But it's not obvious how quinoa can stop being a superfood and start being a staple again. Quinoa's status as a fashionable object of desire in the West has pushed the whole market well beyond the economic reach of the hungry of the world. Now that quinoa has become a luxury good, it won't be easy to transform it back into a subsistence crop for the people who really need it.

Quinoa is not the only recent food trend to have generated unintended consequences at the point of production. From 2006 to 2016, the amount of avocado eaten in the United States more than quadrupled, to over a million tons a year. To feed that demand, between 2001 and 2010, avocado production tripled in the Michoacán region of Mexico, leading to deforestation and excessive water use. Hectare for hectare, an avocado orchard uses more than twice as much water as forest, meaning that forest wildlife in the region has been doubly hit: first by the loss of forest and then by the depletion of mountain springs. The growth of the Mexican avocado industry has been an economic boon. In 2016, more than eight out of ten avocados exported worldwide came from Michoacán. But Guillermo Haro, the attorney general for environmental protection in Mexico, urged caution. The country's forests are, he has said, a "wealth greater than any export of avocados."[19]

In the context of Western café life, few options feel more benign than avocado toast. It's what you choose when you are being "flexitarian" and trying not to eat bacon. It's what you have when you are being "good" and avoiding cake. It's what you eat because you love the jade-green color and the lush texture of the avocado, contrasting with the crispness of the toast. It's what you eat because half the people in the café are eating it and it would feel strange not to join in.

Yet the avocado trend in the West (and in China, where avocado imports are growing 200 percent every year) has made the life of many Mexican avocado farmers both more profitable and much more dangerous. In the 2010s, the swelling profits to be made from avocados in Mexico attracted the attention of local drug cartels, who imposed "taxes" on avocado growers. Those who refused to pay might have their family members abducted or attacked or have their farms burned down. This crime wave has led some to allude to Mexican avocado as "blood guacamole."[20]

No one who eats quinoa or avocado would ever have wished their new tastes to have such violent consequences. But this is the trouble with food trends. When patterns of consumption change so capriciously across the planet, there is no time to take stock and consider what the effects of our changing tastes will be. When we buy the latest trendy health food, we do not anticipate that the people who produce it will suffer. Nor do we anticipate that the new food that we desire so much may be a fake.

FADS AND FRAUDS

If you want to choose a food or drink that is particularly likely to be adulterated or mislabeled, a good place to start would be anything trendy. "Food fraud and food fads go together hand in glove," says Professor Chris Elliott.

Elliott, who speaks in a reassuring Northern Irish brogue, is one of the world's leading experts on food fraud: how it happens and how it can be stopped. In the wake of the "horsegate scandal" in the United Kingdom in which it was discovered that cheap hamburgers had been padded out with horsemeat, he wrote the Elliott report on the integrity of food supply networks. Elliott is pro-vice-chancellor at Queen's University Belfast, where he runs a state-of-the-art lab with the capacity to analyze the authenticity, or lack thereof, of ingredients. Is this batch of herbs labeled "oregano" really oregano or is it sumac padded with olive leaves? Elliott and his colleagues are constantly scanning the food supplies of the world trying to predict and prevent the next food scandal before it happens. A sign of how little the food fraudsters want Elliott to succeed in his work is that the lab where he analyzes food is fitted with bomb-proof doors.[21]

Food fraud is an old, old phenomenon—as ancient as the buying and selling of food itself. There have always been traders who took expensive saffron and padded it with something cheaper, innkeepers who sold watered-down beer, sellers who sold their produce at false weight. But our new globalized food supply offers opportunities to cheat on a scale that has never been seen before.

Looking at trends is at the heart of what Elliott does. Food fraud happens for a number of reasons—the greed of sellers, the laissez-faire attitude of governments, and long food chains that no single person can oversee—but one of the main causes is a mismatch between supply and demand. When there suddenly isn't enough of a given ingredient to meet demand, that lack creates a strong incentive for fraudsters to step in and sell a fake version.

For this reason, the trendy superfoods that become instant objects of desire may be among the most compromised in the whole food supply, even as they claim to be the purest. Pomegranate juice is a case in point. In the first decade of the new millennium, the popularity of pomegranate products skyrocketed, helped by marketing and health articles proclaiming its supposedly unique antioxidant properties. The United States consumed 75 million eight-ounce servings of pomegranate juice in 2004, but just four years later, in 2008, Americans consumed 450 million servings, a colossal increase.[22] At the same time, consumers of pomegranate juice were becoming more demanding. In 2004, half of all the pomegranate juice that Americans bought was labeled as a blend of pomegranate juice and some other cheaper juice; by 2008, three-quarters of the pomegranate juice sold claimed to be 100 percent pomegranate. This makes sense. If you are buying pomegranate juice for the health benefits, you want to get the full benefit.

There was just one small problem. Data seen by Elliott at the time told him that there weren't enough pomegranate trees in the whole world to be meeting this unprecedented thirst for pomegranates. It takes eight years for a newly planted pomegranate tree to bear fruit. It was as clear as day to Elliott and other food fraud experts that not all of the "100 percent pomegranate" juice being imported into the United States and United Kingdom from Iran, Iraq, Syria, and other Middle Eastern countries could possibly be pure.

The financial incentives for pomegranate juice fraud were huge. In 2013 prices, a gallon of pure pomegranate juice concentrate cost thirty to sixty dollars wholesale, whereas a gallon of apple juice concentrate could sell for as little as five dollars. By cutting pomegranate juice fifty-fifty—or more—with apple or grape juice, producers could make vast profits. The juice would be shipped to a distributor in India or China or Russia, repackaged, and then shipped to the West for bottling, to avert suspicion. The bottler might have had no idea that the pure pomegranate juice he was buying was fraudulent. The consumer—who had never tasted 100 percent pure pomegranate juice—also had no idea that the healthy juice he or she was paying a premium for was diluted with other juices.[23]

When I interviewed him in 2016, Chris Elliott was turning his gaze to coconuts. Since the Dutch tulip bubble of 1637, few products have attracted such sudden and disproportionate interest as coconut water. This bland drink is perhaps the trend of all trends at the moment: a supposedly uniquely "hydrating" beverage that sells for high prices as consumers search for an alternative to sugary soft drinks. The global coconut water market has been forecast to grow at a compound rate of 26.7 percent until 2020. It looks like a prime target for fraud. "The easiest way," says Elliott, "would just be to water it down and add sugar." The nutritional profile of coconut water varies depending on the time of year and the harvest, so if a batch happened to show up watery on a profiling test, it could be put down to natural variation.[24]

As with pomegranates, coconut products would be a gift to fraudsters because health-conscious consumers will currently pay almost anything for coconut-based foods, ranging from coconut sugar to coconut vinegar (yes, such a thing exists). For years, coconut was seen as something to be avoided on health grounds because of its high levels of saturated fat, but it is now revered by fans of clean eating as a "good fat" (although the health benefits of coconut remain contested).[25]

In 2016, the coconut harvest in Thailand failed, leading to a sudden shortfall in supply, and some of the world's leading processors of coconut water, such as PepsiCo, moved their operations to India. Elliott can't say for sure yet how much coconut fraud, if any, is happening. But it does look like a perfect storm for swindling: a premium health product for which there is an insatiable demand and whose supply is becoming erratic. Like pomegranates, coconut is not something you can produce more of on a whim. If a pomegranate tree reaches fruition slowly at eight years, a coconut palm is even slower: ten years.

Chris Elliott, who is in his midfifties, remembers a time when food habits were slower and steadier. He grew up on a small farm in County Antrim: "A few cows, some potatoes, and soft fruit." Food trends were out of the question for Elliott's family. In his whole childhood, he never once tasted a cumin seed. The only new foods they ate came via bartering with neighboring farms. Sometimes when Elliott was a boy, his grandfather went off with a van full of potatoes and returned with a van full of cabbages. But when Elliott compares his diet then with what he observes most people eating today, he's aware how lucky he was. His family was not rich, but they ate fresh food at every meal, simply cooked, whereas now he finds that "fresh produce is nearly a luxury item for many people."

When he was a child, Chris Elliott knew what he was eating. He worries that such knowledge is becoming a rarity. Many children in the United Kingdom do not know that bacon comes from a pig or that milk comes from a cow. With the rise of healthy food trends, Elliot fears that our disconnect from the realities of food will become even greater. The more avidly we go in search of the latest faddy ingredient to cure our dietary ills, the more we make ourselves vulnerable to food markets that we have no control over.

Where will this pursuit of the new end? And what are the costs? I know I am extremely lucky to be one of those with skyr in my fridge. Sometimes I think about how limited the food options were for people of my grandmother's generation, and I marvel at the sheer options of global flavors available to me now: the licorice scent of Thai holy basil, the mysterious flavor of pandan leaf. Other times I think about the old foods that we abandoned through no fault of their own but because they weren't quite new or exciting enough. Every pot of skyr that gets eaten is a rice pudding that never saw the light of day.

I also worry about the eaters who are left behind. In all her years of analyzing food trends for Mintel, Lynn Dornblaser has been struck by the fact that low-income consumers do not participate in trends to the same extent that those on higher incomes do. Food trends can become yet another form of cultural exclusion. Dornblaser remarks that, in the United States, consumers earning $50,000 a year or less "are never at the leading end of a trend. Ever. As a low-income consumer you can't afford to try something and have it get wasted because your family didn't like it." As we have seen, having the money to participate in the pomegranate or coconut trends may be a mixed blessing, but at least you have the chance to join in with the ridiculous fad if you want to. All her years of studying consumers tells Dornblaser that low-income shoppers have exactly the same aspirations to buy good-quality food that anyone else does. The rise of expensive trendy foods widens the gap between affluent eaters and everyone else because it reinforces the idea that good food is an exclusive pursuit.

Many of the most hyped new foods of recent years are simply those with the largest marketing budgets. Professor of nutrition Marion Nestle has exposed the fact that the food industry pays scientists and bloggers to promote particular foods at the expense of others. When we read an article saying that the pistachio nut is the healthiest of all nuts, the odds are that the information and research in it has been paid for by the pistachio industry. The marketing of expensive superfoods masks the important truth that a healthy diet need not include a single fashionable ingredient. No superfood can equal the value we would get from consuming a daily mix of vegetables and fruits, plus a range of other wholesome foods—but this is the kind of boring old news that no one wants to hear these days.[26]

Food does not need to be reinvented every five minutes, any more than water or air does. Perhaps the biggest problem with our current penchant for culinary trends is that they are a sideshow distracting us from more fundamental concerns—such as how to combine the busyness of modern life with steady meals that give us both pleasure and health. There is something unmoored and manic in the way we grasp at novelties as the solution to our diet ills while neglecting the basics. This is the behavior of a generation who has lived through so many changes to our diets that we sometimes seem to have forgotten what food actually is.

CHAPTER SIX

DINNER WITHOUT DUTY

WALK THROUGH THE CENTER OF ANY CITY OF THE world around 7:00 p.m. (or 9:00 p.m. in the Spanish-speaking world) on a summer night and you will see hundreds of people eating out, in countless different ways. On outside tables in a bustling square, couples are sipping beers or Negronis and twizzling spaghetti on forks. Big groups of twentysomethings are sharing wood-fired pizzas and laughter. A family with young children is eating Lebanese mezze, offering little bites of falafel to a toddler. Lovers of Asian food are eating sushi and fiery, sinus-clearing Vietnamese noodle soup.

There are solo diners too. You see a single person perched on a corner table of a French-style bistro with a plate of hearty cassoulet, a glass of red wine, and a book, having a night off from cooking. I want to be that person, all dressed up and eating something delicious cooked by someone else, in a room with more stylish lighting than my own kitchen.

At a restaurant, you feel you are at the heart of things but also that you can disappear into the night anytime you want. You are free to explore new cuisines and dishes that you might never dare to cook at home or to order comforts you were denied as a child. A modern restaurant meal is the opposite of a family dinner table in the old days, where everything was overseen and controlled by a parent. This is food with all the judgments and obstacles taken away. Never before have so many people had access to the free and easy pleasures of eating somewhere other than home. Around half of all the money Americans spend on food now goes on meals out.[1]

Since the start of the new millennium, eating out has become not just our preferred way of eating but our preferred form of entertainment, full stop, especially among the young. This is a sign that, despite our time-pressed lives, we still value meals and think of them as something worth spending money on. Toby Clark, who works for the consumer research company Mintel, has attributed the growth in spending on eating out to a new desire in consumers to spend money on experiences "rather than accumulating more things."[2]

The current excitement about eating out for fun is partly that we have come so late to it, at least in its current form. Since ancient times, there have been cookhouses and bake houses, where customers could buy tasty meals to take away. But to sit and be served hot food for pleasure rather than need was a relatively rare luxury, except for travelers who ate at roadside inns. Now, for millions of people, the joy of meals out has become an everyday extravagance as well as a form of self-expression. Shall we have Korean barbecue tonight, or would you rather try that new Sicilian place 'round the corner?

We are eating out more than we ever used to, and these meals away from home are an edible sign that more people have disposable income than in the past. The colossal surge in eating out since the 1990s—which is a global phenomenon—could never have happened were it not for the happy fact that large numbers of people are richer than ever before. In South Africa, average incomes nearly doubled between 1994 and 2010 (from $3,610 to $6,090), and over the same period, there was a vast increase in the number of inexpensive restaurant chains in the country, from Nando's chicken to Debonairs Pizza.[3]

Across the world, there has been a stupendous rise of eating out, with restaurant meals transformed from a once-a-year treat to something so casual you might have it on a whim in a shopping mall. This is part of a bigger story, which is about the way we have been liberated from so many of the old social duties around food. No one has to cook, if he or she doesn't want to, and the freedom with which we obtain our food now is completely at odds with the way we shopped for food and ate in the past. We can afford to give our appetites free rein in a giant marketplace of flavors.

Whether we are eating out or shopping online, eaters are now able to gather a smorgasbord of food with fantastic ease and without any of the old obligations to consume it in traditional ways. "Never could we have imagined that we would ever have this much to eat," commented a Chinese grandfather in a 2007 interview with writer Faih d'Aluisio. After living through decades of scarcity, this grandfather described his unease at the sight of young children buying strips of fried dough to eat for breakfast on the way to school and throwing half away uneaten.[4]

These changes in how we buy our food are not just about eating but about the way we organize our lives. In today's world, much of the time we eat our food in public and we shop for ingredients online in private. It used to be the opposite way around. Food shopping used to be public, whether it was done at open-air markets or small independent grocers' shops. It was impossible to buy any item of food without being exposed to the counsel and judgment of others. This was how many of the rules of cuisine, and of behavior around food, were maintained. You bought the same foods that you saw other shoppers buying. Conversely, eating was mostly a private affair, conducted at the family dinner table, where parents dictated what we ate, and how much, and how we should feel about particular ingredients.

No one seems to oversee or obstruct our food choices anymore. As with other aspects of the nutrition transition, these transformations in how we get our food have produced both good effects and bad ones. This age of cafés and convenience has brought extraordinary new freedoms with respect to food, but not everyone has shared in these freedoms to the same extent. What's more, the freedom to eat whatever we choose is not always quite as liberating as it seems. What we have gained in social and economic freedom, we have often lost in health. Although they have made our lives easier, the growth of the restaurant and the supermarket are also a major factor in why so many are suffering from diet-related illnesses. Food was never meant to be quite this easy to get. The sheer overload of dishes we are offered at any given time in any given metropolis is out of all proportion to anything our brains and bodies have known before. It isn't always easy to judge when to stop eating when all around us there are bistros and takeaways and cafés and gastropubs sweetly calling us in.

THE DEMOCRATIC RESTAURANT

Eating out has become so normal that it can be hard to remember that there was a time when it wasn't. For thousands of years, home-prepared food was the source of most of the calories eaten in the world. In America, as recently as the 1950s, a restaurant meal, taken for pleasure, would have been a major undertaking for an ordinary family.[5]

Except for the moneyed classes, most eating out tended to be more about utility than fun. In the United States, there were casual places such as diners or lunch counters, aimed at working people, where you could get homely short-order food such as sandwiches or eggs and hash browns and endlessly replenished cups of watery coffee. But it was rare to go out for a dinner with tablecloths and wine and food that was different from the food of home. What was true of the United States was true of almost everywhere in the world (except for certain countries such as France and Italy, which have long had a tradition of small and inexpensive family-owned eating places). As of 1960, a survey found that 84 percent of people in the Netherlands ate out "rarely or never." To be a frequent restaurant-goer then was the

preserve of the very rich. Restaurants in the Netherlands had a reputation for being too stuffy, too formal, and too expensive.[6]

Just as food shops of the past could be places where customers
felt judged, the restaurants of the past offered expensive delights from
which most people felt excluded. In Britain, restaurants had a reputation for being snobbish places where toffs took champagne and
oysters while the rest of us pressed our noses to the window, out in the
cold. They felt starchy and fussy, places where you might be judged
harshly for using the wrong fork.

In the space of just a few decades, eating out for pleasure stopped
being exclusive. By 1980, the number of Dutch people who ate out
"rarely or never" were in a small minority: just 26 percent. The same
trend could be seen in other countries. In 1959, the average proportion of a British person's food budget spent on eating out was just 9.6
percent. By 1995, the proportion of food money spent on eating out
in Britain was 28.4 percent, nearly a third. Maybe the most telling
change was that the *average* British person now devoted a similar percentage of his or her food spend to eating out as a very wealthy person.
This was a remarkable shift in such a short space of time. Eating out
has been a great social leveler.[7]

Before eating out could establish itself as a popular habit, several other things had to change. There had to be enough people rich
enough to buy restaurant food; there needed to be enough modestly
priced restaurants—and other eateries—to serve them; and customer
tastes had to change enough that we actually felt it was worth paying
more to eat this new food.

It takes a certain confidence to spend money on eating out, the
confidence that comes from knowing that your budget has wiggle
room. The first change that prepared the ground for more eating out
was the growth in incomes around the world, meaning that people
had spare cash to spend on treating themselves to good meals out.

Eating away from home is not always a sign of wealth, any more
than home-cooked food is always cheap. Sometimes people buy food

away from home because they are too poor to own a kitchen. A case in point is street food in Asia, which is often bought from cheap stalls and carried home in plastic bags: maybe some fried rice with scraps of salted fish or a green curry with roti bread. But most forms of eating out—even when it's fast food—cost more than the most basic home-cooked meal. In every country, as incomes rise, people eat out more. Like other aspects of stage four, eating out is what happens when countries and people become richer, when more women start to take paid work outside the home, and when workers move into cities.[8]

The second change paving the way for more meals out was that we needed a new type of eating establishment, cheaper and less formal than the old brasseries of the classic French tradition. People would never have enjoyed so many meals out in the late twentieth century had it not been for the new restaurants run by immigrants selling "ethnic" food, from the Indonesian restaurants of the Netherlands and the Vietnamese pho of Seattle, to the Italian pizza and pasta of virtually everywhere on the planet. The success of modern restaurants is a story about immigration and the edible riches we have gained from the migration and interchange of people around the world.

The third and crucial change was that our palates had to develop to the point where we wanted to eat this new food, so unlike the food of home. We had to open our mouths and our minds to unfamiliar flavors.

"They couldn't understand how can a thing be sweet and sour at the same time," Woon Wing Yip recalls of British customers at Chinese restaurants in the 1950s and 1960s. Wing Yip, who was born in 1937, opened his first Chinese restaurant in Britain in the seaside town of Clacton-on-Sea in 1962. He had arrived from Hong Kong in 1959 with just £2 (a few dollars) in his pocket and found work as a waiter. In the early days of his restaurant, he often heard customers making negative comments about the menu. They would stand out-side and say "sweet and sour pork" sarcastically as if such a bizarre combination must be a joke. But despite their initial uncertainty,

British customers soon realized that Chinese restaurants—like Indian curry houses—offered them a level of comfort and service that no British eatery could provide for the same price. By 1970, there were four thousand Chinese restaurants in Britain (compared with just one in 1914).[9]

Like so many other inexpensive restaurants around the world, British Chinese restaurants eased customers into the new tastes by letting them order anything they wanted, including their old home comforts. If a British person happened to fancy eating curry chicken with bread and butter, they could have it without feeling that anyone was snickering at them for being gauche. When I was in India in 2016, I spoke to many Indians who told me of Indian Chinese restaurants that offered them similarly inauthentic pleasures, creating dishes tailored to local tastes that combined the cumin and green chili of India with the soy sauce and vinegar of China.

The early Chinese restaurants in Britain treated customers like princes, whose every wish was right and who deserved to eat a banquet even if the only occasion being celebrated was Thursday night. "We put a carpet down," Wing Yip recalled in an interview about the early days of British Chinese restaurants. "We put a tablecloth down. Before that restaurant, you only got a carpet and a tablecloth and a waiter service in hotels, which were beyond ordinary means and they close half past nine." I will never forget the glamour of being taken to a British Chinese restaurant as a child in the 1980s. Sizzling savory dishes, darkly redolent of soy sauce, arrived on hot plates. The table was laid with smooth white-and-pink tablecloths that became splattered with a Jackson Pollock pattern of different-colored sauces as the meal went on.

Back then, a meal at a restaurant was still a rare and special occasion, but it would become much less so. Spending on eating out increased by 33 percent in the United Kingdom between 1985 and 2005. In the United States, it rose even faster, increasing by 76 percent, so that by 2005 the average American was spending $2,500 a year on eating out.[10]

A meal out with friends is one of the great joys of life. As a greedy person, I sometimes feel a lurch of gratitude to be alive now when there are so many different and delightful ways of eating out, so many ways to be consoled and diverted by food, so many booths and tables to uplift you. In T. S. Eliot's poem, Prufrock measured out his life with coffee spoons. I could measure out mine in restaurant tables, each of which opened up a new world of flavor to me. As a child there were the curry houses where I watched my father turn crimson from the heat of vindaloo; later, in my twenties, I can see myself at a table in Venice eating a seafood risotto as flowing and delicate as water; and, years later, the Spanish restaurant where I gulped down tapas, through tears, after my grandmother died.

BUT THE RISE OF EATING OUT HAS BROUGHT PROBLEMS as well as pleasures. One of the problems is that when special-occasion food becomes a regular occurrence, we don't seem to derive quite the same degree of pleasure from it as before.

A team of British food sociologists conducted two studies on attitudes to eating out, first in 1995 and again in 2015. Back in 1995, the sociologists found that most British people—based on a sample of just over a thousand—considered eating out to be very special. The sample group told the researchers that they looked forward to these experiences immensely. They spoke of being entertained and gratified by every aspect of a restaurant meal: the company, the food, the sense that they were participating in an event.[11] It was as if they were determined to derive satisfaction from the experience.

When the survey was repeated in 2015, the respondents had lost much of that sense of joy. Eating out now happened much more frequently for most people than twenty years earlier, so much so that they had started to take it for granted. Yet the more of it they did, the less they reported enjoying it. Much of the old specialness and sense of occasion had gone.

Neither are meals out necessarily sociable anymore. Over the past ten years, the world's high streets have been flooded with new eating places that are somewhere between a fast-food joint and a restaurant, designed to cater to people who hardly have time to eat. There are juice bars and places that sell beetroot and quinoa salads, and healthy Japanese places selling everything from sushi to salads and miso soup. There are soup cafés and burrito cafés and thousands of Starbucks. It's a luxury to be able to choose between such a variety of different foods and cuisines, even for a normal working lunch, without anyone overseeing your choices. And yet there's also something banal about all this food on the go. The mood of such meals is casual and forgettable.

A more serious problem with the rise of eating out is that it has come at a heavy nutritional cost. One study from the United States found that eating *just one meal* away from home every week translates over a year to a weight gain of around 2.2 pounds, equivalent to 134 calories a day. And most people in the United States—and elsewhere—don't eat just one meal away from home, but many.[12]

CALORIES AND CONVENIENCE

In general, the nutritional quality of what we eat when out is not the same as what we eat at home. Analysis by the USDA in the 1970s found that the nutrient quality of food eaten out in the United States tended to be significantly lower in vitamins and higher in calcium and fat than food eaten at home. Back in the 1970s, this didn't matter for the overall quality of American diets, because eating out was still a rare treat back then. It would have been unwise to fret too much over the calories in a creamy dish of pasta Alfredo when it was a once-a-month indulgence.[13]

It is different now that ever more meals in the week are eaten out. The nutrients that we get—or don't get—from these meals starts to matter more. It isn't just in America that diners' health is being affected by eating out. A study of nearly forty thousand people across ten European countries found that eating out was associated with consuming more calories and eating more sweet foods than eating at home. This makes sense. When we go out for a meal, we feel we are treating ourselves, so we want ample portions and dishes that comfort us. Meanwhile, restaurants themselves know that if they don't supply us with enough of these comforts, we may take our custom elsewhere.[14]

There is no intrinsic reason why restaurant food shouldn't be every bit as healthy as home food. But many Western restaurants—whether fast food ones or not—perpetuate the idea that wholesome items such as vegetables can never be something to relish. Research from the Food Foundation in Britain in 2017 found that for every three restaurant meals eaten—not just fast food—the average restaurant diner in the United Kingdom ate just half a portion of vegetables.

When eating out with my children, I often order a couple of dishes of green vegetables on the side—because we love them—and the waiter or waitress will sometimes express a kind of bantering astonishment that we are being so healthy. The experience of ordering greens in a Chinese restaurant is quite different, we have found. Instead of querying our desire for vegetables, the waiter is more likely to advise us on whether to choose bok choy or gai lan and the merits of garlic versus oyster sauce.[15]

Every human life is peppered with sadness and loss, and there are far worse things than drowning our sorrows—or celebrating those rare moments when everything goes right—in a lovely meal out: to sit at a table that isn't your responsibility, eating food that someone else has labored over. "A man hath no better thing under the sun, than to eat, and drink, and to be merry," as the Bible has it. The difference now is that the feasting is so frequent that it has itself become a cause of our maladies.

If an abundance of meals out can damage our health, that is partly because so much of our eating out now consists of fast food. Around half of all restaurants in the United States are now fast-food restaurants, which offer an even more intense feeling of eating without duty than a full-service restaurant. Fast-food franchises have traveled the world, and wherever they have arrived, they have been rejected by some but welcomed by others as the taste of modernity.[16]

I've read *Fast Food Nation*, Eric Schlosser's 2002 exposé of the true forces behind a fast-food burger and fries, which revealed the horror of giant slaughterhouses in which some of the ground beef becomes contaminated with feces. I've seen *Supersize Me*, Morgan Spurlock's 2004 documentary in which Spurlock ate only McDonald's food for thirty days and gained twenty-four pounds, mood swings, and a fatty liver.

And yet almost nothing I have ever eaten tasted as good as the Big Macs I had on a trip to Moscow in 1993, a couple of years after the fall of the Soviet Union, when I was briefly staying in a cockroach-infested student apartment. Most Russian food shops at that time were still stuck in the mindset of the communist era. You might queue up to buy a loaf of black bread in a shop with half-empty shelves. But at the McDonald's in Pushkin Square—which opened in January 1990—there was an air of plenty and delight, and the place was full of people eating exactly what they wanted, so long as what they wanted was burgers and fries and chicken nuggets and thick icy-cold milkshakes.

Sometimes fast food is snobbishly looked down on by people who have never heard the siren call of a ninety-nine-cent hamburger. The assumption is sometimes made that these places—because they cater to a poorer clientele—offer a lesser degree of pleasure than a sit-down restaurant meal. In my experience, this is far from true. The pleasures of fast food—of being in a room that isn't home, eating strongly seasoned food that you haven't cooked—can be every bit as intense as the emotions you may feel at a ten-course tasting menu at a Michelin-starred restaurant. The difference is that many fast-food customers—if they are heavy users—pay for their pleasures with a series of chronic health problems.

There's now enough evidence to say beyond doubt that regular and frequent consumption of fast food will increase a person's risk of heart disease, as well as the risk of developing insulin resistance and type 2 diabetes. This is perhaps not surprising given that fast-food meals tend to contain more fat and sugar than home-cooked meals, as well as lower amounts of vegetables, fiber, vitamins, and milk. By 2007–2008, a typical American adult who ate fast food was getting 877 calories a day from these meals.[17]

This is not to say that every devotee of fast food will gain weight or become unwell. As always with food, it's complicated. One of the few studies to separate out the risk factor from different types of fast food found that, in a cohort of forty thousand African American women over the age of thirty, the highest rates of type 2 diabetes after ten years were among those who ate burgers and fried chicken more than twice a week, with more modest risks of diabetes from regular take-away meals of fried fish or Chinese food, and no increase in diabetes risk at all from Mexican food or pizza. Make of that what you will.[18]

Fast food is one of the mechanisms through which health inequalities play out because there tend to be higher concentrations of fast-food outlets in poorer neighborhoods. There is clear evidence linking the availability of fast-food restaurants per capita and the rate of childhood obesity. This is no small problem, given that one in three American children now eats fast food every day, a rate that has increased fivefold since the 1970s.[19]

The nearer you live to fast-food restaurants, whether as an adult or a child, the more likely you are to be obese. One study from 2010 looked at the weight outcomes of living near a fast-food restaurant for three million children and three million pregnant women in the United States. The researchers, led by economist Janet Currie of Columbia University, found that having a fast-food restaurant within 0.1 miles of a school would result in a 5.2 percent increase in child obesity rates.[20] With its large sample size, this study was the closest any research has come to demonstrating that proximity to fast-food restaurants can actually cause obesity (rather than just correlating with it). The study wasn't perfect, however. As Currie pointed out, ideally she would have also studied people living in American neighborhoods where fast food is scarce, as a point of comparison. But this study would be impossible to set up, because there is nowhere in America where fast food is scarce.[21]

How do we define fast food? Most of us know fast food when we see it, yet there is little consensus on what it really is. The *American Heritage Dictionary* defines it as "inexpensive food, such as hamburgers and fried chicken, prepared and served quickly." Other definitions mention that fast food tends to have a limited menu and no table service and that the food comes in disposable wrappings or containers. But on this definition, there is no difference between "fast food" and an independent food truck selling homemade vegan tacos. Currie and her colleagues constructed several measures of fast food but found that the best benchmark definition was simply to focus on the top ten chains in the United States, which in 2010 were "McDonald's, Subway, Burger King, Pizza Hut, Jack in the Box, Kentucky Fried Chicken, Taco Bell, Domino's Pizza, Wendy's, and Little Caesar's." These chains are notable not only for the fact that they tend to sell inflated portions of food high in refined oils, sugars, and carbohydrates but for the aggressive ways in which this food is branded, marketed, and advertised.

The marketing of fast food can induce people to buy it even when we recognize it is worse in nutrition and flavor than more traditional foods. In Thailand in 2009, a team of researchers interviewed more than six hundred teenagers, asking them how they felt about fast food. More than three-quarters of these Thai teenagers knew that fast food such as fried chicken or burgers was high in calories and that it was more likely to cause obesity than traditional Thai dishes fresh with green herbs and galangal. Yet more than half regularly ate Western fast food for reasons that went beyond its flavor or texture. They loved fast-food meals because they saw them as modern. They liked the marketing giveaways and price promotions and the fact that they recognized the food from TV advertising. Above all, they loved the sheer quickness and the efficiency of grabbing a burger and a fizzy drink to go. This food might not be as healthy or even as tasty as the food they grew up eating, but what it offered was total convenience.

Fast food has changed the way we all eat, whether you ever set foot in McDonald's or not, because many consumers are now starting to demand a fast-food level of convenience in everything that we eat. Many of us now seem to want food that makes us feel like wizards, magicking meals out of thin air.

DINNER ON A BIKE

"The weird part is you never see the boss. The boss is just a nameless faceless app." It's the summer of 2017, and I am talking to Zack, a British eighteen-year-old, about the part-time job he does alongside revising for his final-year exams at sixth form college in Cambridge, the university city where we both live. Zack is one of more than thirty thousand delivery riders worldwide working for Deliveroo, a start-up founded in February 2013 that delivers hot restaurant food—by cycle or scooter—minutes after it has been ordered on a smartphone. As of September 2017, Deliveroo had a valuation of more than $2 billion and was operating in 150 cities covering twelve countries.[22]

Only recently, it would have seemed like an unimaginable luxury to ask someone to bring you a steak frites with béarnaise sauce on a bike. Now, at least for people who live in cities, it's utterly normal. On any given evening in Cambridge, there are so many Deliveroo riders on the road that I sometimes have to pinch myself to remember that until a couple of years ago, there were none.

They look like blue cycling tortoises weighed down with their backpacks full of pad thai and pork ribs and many, many portions of fries. Every Friday afternoon, Zack fortifies himself with a big meal of pasta and a heated-up frozen chicken burger, straps on a vast turquoise-colored insulated food bag, and gets on his bike for a six-hour shift. He cycles all over the city, pushing himself to go as fast as he can because he is paid extra for every delivery ("My parents worry about me on the roads," he says). On a typical Friday-night shift, he may deliver anything from a Chinese banquet for eight people to a single-portion Mexican meal for a student too preoccupied with an essay to cook. "Once I had to deliver a single burrito and twelve bottles of Corona beer to one guy."

Like many of his teenage friends, Zack is not just a Deliveroo employee but a Deliveroo consumer. Part of why he likes the work, he tells me, is that it gives him spare cash to spend on socializing with his girlfriend on a Saturday night. Often as not, this means staying in with Netflix and ordering dinner through Deliveroo. "We watch reality TV while we eat and then a movie afterwards." Zack, who will soon be off to university to study science, jokes about the fact that his bank statements "basically say Deliveroo in or Deliveroo out."

In an incredibly short space of time, Deliveroo—and American versions such as Seamless and UberEATS—has revolutionized how many people get their food. It is the ultimate in dinner without duty. For those with the app and the money, these delivery services take a whole city's worth of restaurants and make them instantly accessible from a sofa, an office, a student hall of residence—anywhere. It's not that food delivery in itself is new, but once it was limited to certain foods, notably pizza, which has been delivered by Domino's and others since the 1960s. In 2016 alone, US consumers spent $10 billion on pizza delivery. What's new about Seamless and Deliveroo is that they take the convenience of delivery pizza and apply it to every genre of restaurant, from French bistros to Japanese sushi bars, from big franchises to small independents, and to every meal of the day. By 2016, delivery orders made up around 7 percent of US restaurant sales.[23]

For some family-owned restaurants, these delivery services have been a lifeline. I spoke to the owner of a small chain of Indian thali restaurants in Bristol in the United Kingdom who said that his turnover had increased exponentially since he decided to hook up with

Deliveroo, because the app could reach customers who had never set foot in one of his restaurants. On the other hand, some restaurants say the delivery model has been an economic disaster that is squeezing the soul out of the business because the delivery apps take such a large percentage of revenue—around 20–40 percent on a typical order. With a delivery order, the restaurant loses the chance to make money from the customer ordering drinks. Since the rise of the delivery app, once-bustling restaurants sit half empty at lunchtimes.[24]

These meal apps are changing our expectations of dining out, because suddenly the restaurant can come to you, which removes so many of the obstacles or inhibitions someone might have felt about eating out in the past. It's like never-ending room service. The youngest of three brothers, Zack feels that the way he eats and socializes on weekends is very different from how his two brothers were eating at the same age, in the distant past of the early 2010s. The next brother up in Zack's family is just four years older, but they seem to belong to different generations when it comes to food. His brother spent his weekends partying or clubbing and would buy food, almost as an afterthought, on the way home. Zack feels that the whole focus of his leisure time is more about food and less about going out. "I don't know what I would do with my weekends if I didn't have Deliveroo."

In a food system already fixated with convenience, Deliveroo takes convenience to new extremes. The first time my family of five tried it—we had a Japanese-style meal from the Wagamama chain—I couldn't quite believe that it was so easy to summon up a whole meal, with different dishes for each person. Each portion of udon noodles or ramen soup was portioned up in a vast, sleek, black-lidded plastic bowl. Zack confirms that the packaging involved in all Deliveroo meals is "colossal." He relishes the challenge of fitting the various boxes into his backpack without leakage: "It's like playing Tetris."

Many of us have started eating as if no one is watching. In the past, the gathering of food was a deeply social act. Seamless and Deliveroo are the opposite. It's a process that seems to make its customers feel detached from normal social obligations, maybe because the food has been summoned up impersonally using a debit card on a touch screen. One Deliveroo courier in London told *The Guardian* in 2016 that he found it "alienating" to be asked to pick up a single Nutella crepe from a restaurant "and deliver it right to someone's desk in the City of London. At times like that you think: what am I doing? This is like late capitalism." This courier also commented that when they get the delivery, "mostly people just take their food and don't say anything." Zack tells me that he went through a "40-delivery dry streak" of no tips. When people do tip, in his experience, it will usually only be around 50–60 pence (60–70 cents). For student workers like Zack who do a few shifts for extra cash, this lack of tipping is no big deal, but many delivery couriers are working full time, with families to support. They are part of the new "gig economy" in which millions of workers find themselves doing jobs with no safety nets or benefits to fit the brutal vagaries of "on-demand" businesses.[25]

Deliveroo is the end point of a much bigger series of transformations in the way that we go about gathering food in the modern world. The buying and selling of food used to involve a series of daily encounters with other human beings. But now, it often involves only interacting with a series of brands using a few clicks on a computer, no eye contact required. The act of food shopping, which used to be deeply public and social, has become steadily more anonymous and private. With online grocery shopping, you can make your food choices without feeling that anyone is watching or passing judgment. To our ancestors, that would have been an unimaginable liberation, as well as quite shocking.

SELF-SERVICE

Of all the reasons that we first fell in love with supermarkets—the ease, the choice, the fact that no produce ever goes out of season, the illusion that it's saving us time, the thrill of pushing a trolley slightly too fast—we sometimes forget that some of the biggest changes they wrought in our lives were social. Supermarkets gave us permission to pick things up and help ourselves, without having to ask a shopkeeper or market stall holder. They allowed us to shop for food in a state of privacy for the first time, watched only by the packets themselves. Or so we thought.

The shift from a world of small independent food stores to one of large, impersonal retailers has been dramatic and ubiquitous. In 1956, there were only 13 food shops in the whole of Puerto Rico that could be classified as a *supermercado*, or supermarket. By 1998, the number of *supermercados* in Puerto Rico had increased to 441. The same thing has happened almost everywhere as huge retailers such as Walmart expand across the globe. In the United Kingdom, just four supermarkets control 75 percent of all food purchased.[26]

Even more than a restaurant, a supermarket gives us the illusion of buying food in a state of freedom. Yet just as a menu will push our choices in certain directions with clever wording, the architecture of a supermarket will shape our decisions in aggressive but largely hidden ways. Through special offers and positioning on the shelf, a retailer can encourage us to put things in our trolleys that we never intended. In the supermarket, without a living human shopkeeper to talk to the customers, those bright boxes and jars have become like "silent salesmen." It was in the supermarket that we started judging food, not on its own merits, but on the font chosen by a graphic designer and the garish promises of a marketing team.

We speak of making "good" and "bad" food choices, but many of the choices have already been made for us the minute we step inside a supermarket. There's gathering evidence that where we buy our food can have a significant impact on our diets and health. Low-income consumers are often hampered by living in food deserts, more than a mile from the nearest supermarket or large grocery store. As of 2011, 23.5 million Americans lived in food deserts. But if there are food deserts, there are also food swamps. Many low-income Americans live in neighborhoods saturated with options of places to buy food, but all the affordable options are unhealthy ones. Very small supermarkets that sell no vegetables are bad for our health, but very large supermarkets are also bad, because they encourage us to buy more than we need. Studies from France and Australia suggest that the bigger the supermarket, the greater the prevalence of obesity among its regular shoppers.[27]

Even if we feel that we are free as we walk around a supermarket selecting our groceries, we arrive in the store primed by marketing to feel more favorable toward certain foods. Certain segments of the population are primed by marketing more forcibly than others. In the United States, food marketing has a strong racial slant. Research shows that Hispanic and African American children will end up seeing twice as many ads for confectionery and soda as white children. The food brands know that TV commercials for sodas and cereals featuring black and Hispanic actors can have a disproportionate effect on brand loyalty, because it is relatively rare for these communities to see themselves featured in commercials in the mainstream white-dominated media.[28]

The entire layout of the average supermarket is set up in a way that normalizes the consumption—and overconsumption—of snack foods high in sugar and fat. The vast majority of supermarket price promotions have been for highly processed foods rather than fresh produce. In China, 80 percent of all packaged and processed foods purchased are bought in supermarkets. This is partly because the profits in retail are surprisingly tiny—often just 1 or 2 percent. Fresh produce is more of a risk for the retailer because if it isn't sold quickly it will go off and be wasted—a process that is known in the trade as "shrinkage." There's a lot less shrinkage in sugary breakfast cereal than there is in lettuce because of its longer shelf life.[29]

The great question is why we allowed supermarkets to take such a large share of the money we spend on food. Many of us claim to be nostalgic for a simpler and more human form of food shopping.

Such is the appeal of the farmer's market (or the small greengrocer or butcher or delicatessen), where you can once again look someone in the eye as you buy their produce. The tomatoes are naked to your gaze, not hidden away under plastic. You can ask the sellers questions: Do you recommend the clementines this week or the satsumas? How do I cook this piece of fish? Shopping used to be one of the main everyday forms of human interaction, and we lose more than food when we stop talking to the person who sells our daily bread. This way of shopping at fresh food markets is still a way of life in much of Italy and Spain.

But we may as well be honest about the fact that in the days when all food shopping was personal in this way, there were downsides as well as upsides. The personal relationship between salesman and buyer—which we idealize in its absence—could be claustrophobic, snobbish, and uneasy (as well as inconvenient and slow, as shoppers traipsed from butcher to baker to greengrocer). In his memoir of a 1950s childhood, the writer Jonathan Meades recalled shopping in an old-fashioned British grocer, where everyone queued up to be served by shopkeepers in "long tan warehouse coats." Meades recalls the tiny space of the grocer being "so close you smelled your ripe neighbour. No gossipy whisper went unheard." In the days before self-service, customers could feel uncomfortably scrutinized by shopkeepers, as well as by fellow customers.[30]

In a life beholden to shopkeepers, the first forays into self-service felt to many like liberation. There was no one behind a counter to come between you and your desires. The French word for it is *libre-service*: free service. In her history of modern shopping, *Carried Away*, Rachel Bowlby writes of the self-service customer as someone "free to make up her mind between different products rather than be persuaded by someone else. Prices and weights are clearly marked; there is no worry about inaccurate scales or short change."[31]

Self-service food stores probably began in America with Piggly Wiggly, founded in Tennessee in 1916. These early self-service stores were sometimes called "groceterias" because they were seen as similar to cafeterias. But the supermarket only came about in 1930, with King Kullen in Queens's Jamaica neighborhood. Its owner, Michael Kullen, had the idea of combining self-service with a store that offered people all the food they needed under one roof in a former parking lot: everything from bags of sugar to fresh produce, fresh fish, and dairy.[32]

Prior to the 1930s, shop owners had believed that customers enjoyed the experience of being waited on and would not stand for the inconvenience of having to fetch goods for themselves. But it soon became clear that many customers actually loved serving themselves without any intermediary. In 1935, Carl Dipman, editor of the trade journal *Progressive Grocer*, wrote about the benefits of self-service for shifting more food. "The properly arranged store . . . has no unnecessary barriers. It lets women and merchandise meet."[33]

We colluded in the rise of the supermarket because it made us feel so free. Supermarkets created the impression that shoppers were children, alone in a candy store, with no one to chastise them for their choices. No one would raise an eyebrow if you bought an extra bottle of wine or another tub of ice cream. No sales clerk would make you feel miserly for choosing the cheaper option. The only witness to your purchases was the checkout worker. Research suggests that we tend to buy different foods in supermarkets than in other, more traditional food markets. A study from urban Kenya in 2018 found that the rapid spread of supermarkets is encouraging people to buy more ultra-processed foods and fewer nonprocessed whole foods.[34] It's hard to disentangle how much of this is due to the supermarkets pushing us to buy these foods and how much it's a result of changes in our own demands and desires.

When we feel that no one is watching, we behave differently around food. I sometimes imagine how I would eat if I had to declare each of my preferences out loud, like a shopper in an old-fashioned grocer's shop. Mostly, I suspect that shopping like this might curb my tendency to overstock my fridge and cupboards. On the occasions that I go to a local Italian deli to buy cheese and cured meats, I often get embarrassed halfway through by the greedy amounts that I am buying and then stop. In a supermarket, this sense of embarrassment never arrives, and often as not, I end up with too much food.

Supermarkets offer us levels of choice that simply make no sense at a human level. When King Kullen opened in Queens in 1930, it stocked just two hundred items. By the 1990s, the average American supermarket contained seven thousand items. But the biggest leap was between the 1990s and the present day, when an out-of-town hypermarket may offer forty thousand to fifty thousand products.[35]

Our apparent freedom of choice when shopping for food has now gone a step further with the rise of online shopping. Clicking on a laptop in the confines of your own bedroom, you may feel that you can choose exactly what you like, even if the foods that pop up at the top of your search have actually been aggressively edited. Of course, you are still being watched, more so than ever, by the vast data-processing machines that register and track your every choice. But many of us are hardly aware that this surveillance is even happening.

The choice of food products now available to us, whether online or in person, is mind-blowing. As recently as 1970, the British cookery writer Elizabeth David complained that "green root ginger, so important in Chinese, Malay and Indian cooking, is hard to come by." Now, you will find fresh ginger in the most bog-standard supermarket. I remember a time when, if I wanted certain specialist cooking ingredients such as leaf gelatine or maple extract, I would need to travel to a food store in a different town to buy them. Now, all these and much, much more are available online. No foodie gratification need ever be deferred. We have a level of food choice that often feels wonderful but is crazy in its superfluity.[36]

It's also a level of choice that is only available to some. One of the worst contradictions in the baffling food world we now inhabit is the gulf between eaters who have too many food choices and those who have almost none. As our expectations of food become ever fancier and more elaborate, the gulf between eaters grows. As we saw in chapter 3, it used to be that everyone in a given country relied on roughly the same staple food—whether rice or bread or corn—while rich people could afford more trimmings to eat alongside the staple. Now, our concept of a "good diet" has become all trimmings and no staple. One way to eat well on a budget—assuming you have a kitchen—is to cook and eat delicious grains and legumes: dal and rice, lentil stew and bulgur wheat. But many modern diet gurus reject such meals as carb heavy, preaching the necessity of exorbitant green juices, wild meats, and dairy-free yogurt. We live in cities where the options are apparently endless, yet those dangled in front of us can be seized only by some. This problem of social inequality goes beyond food. As the journalist John Lanchester wrote in 2018, "Everywhere, more than ever before in human history, people are surrounded by images of a life they are told they should want, yet know they can't afford."[37]

THE INEQUALITIES OF CHOICE

Modern urban hunger doesn't taste of bread and gruel, as it did in Victorian times. It tastes of canned tomatoes. On a bleak and rainy day a couple of weeks before Christmas 2017, I meet Jonathan Ede, director of Cambridge Food Bank. Cambridge is one of the richest cities in Britain, but it also has some of the highest levels of economic inequality in the United Kingdom.[38] This is a city where poverty hides in plain view. If you are in the Deliveroo-ordering classes, it's easy to close your eyes to what life might be like for people living on benefits on the city's housing estates.

Since the financial crisis of 2007–2008, there has been a rise around the world in the number of individuals seeking help for hunger relief from charitable food banks, or food pantries as they are called in the United States. Across Europe, one of the marked food trends of recent decades has been the rise of food banks handing out free or subsidized food in big cities in response to the hunger that persists in even the most affluent societies. The details of how food banks are set up vary from country to country—and even from city to city. Some food banks depend on donations from individuals; others salvage excess food from farms or supermarkets or manufacturers. Some food banks are run on a warehouse model, passing the raw ingredients that they collect on to soup kitchens. Others—like Cambridge Food Bank and most of the food banks in the United Kingdom—hand out parcels or boxes of ingredients direct to the hungry. Whatever the setup, the basic rationale for a food bank is always the same—which is that, even in affluent, developed cities full of restaurants and cafés and delis and pizzerias, there are huge numbers of families who cannot reliably muster the money to feed themselves.

In Germany, the richest country in Europe, food pantries are now providing free food support to as many 1.5 million people. Those queuing at the Berlin food pantries include single parents and people on welfare, low wage earners and elderly people on stretched pensions, children, and asylum seekers. Many are not sharing in Germany's prosperity, even those in full-time work. Nearly a fifth of the population says that it is worried about poverty, and around 5.6 percent of Germans are officially classified as "poor."

What is happening in Germany is being replicated in all the big cities of the world. "Food banks are extraordinarily new," as Joanna Biggs wrote in an article on a London food bank in the *London Review of Books* in 2013. The number of people turning to food banks in Britain rose from 70,000 in 2011 to 347,000 in 2013. They are the most visible symbol of the stark inequalities in the way we eat now, where some struggle to afford the basic ingredients for a nutritious meal, while others have money to burn on eye-wateringly expensive designer snacks, organic blueberries, and chia seeds. At Kensington and Chelsea Food Bank in one of the poshest parts of London, Bigg witnessed donations of "caviar, loose-leaf Orange Pekoe tea in a suede pouch with tassels and handbag-sized bars of Green and Blacks chocolates." For some of the individuals who donate, food, even luxury food, has become something so affordable and accessible that a jar of caviar becomes mere kitchen clutter. For the individuals who receive, food has become so expensive that even a can of new potatoes may be welcome.[39]

"Some of the people we see are in a desperate state," says Jonathan Ede at Cambridge Food Bank. One woman, whose husband suddenly lost his job, was down to forty pounds (just over fifty dollars) to feed her family of four for a whole month. We stand in the food store and Ede shows me where all the donations are sorted and boxed up. I can see crate after crate of canned tomatoes and bottled tomato pasta sauces, plus whole shelves of dried pasta. A typical box of food for a family of four would include packaged cereal, canned soup, canned tomatoes, canned vegetables, canned ham, cooking oil, a small jar of instant coffee, ultra high-temperature pasteurized (UHT) milk, salt, pepper, canned fruit, pasta sauce, and pasta. Many food bank users can't cook, Ede tells me, but the one thing that everyone can make, in his experience, is pasta with tomato sauce.

In the United Kingdom, this type of food bank has existed only since the year 2000, when the first Trussell Trust food bank was founded in a shed in Salisbury. As of 2017, the Trussell Trust—a Christian charity—was running a network of more than four hundred food banks, handing out emergency food parcels designed to last three days to those who can prove they are in need.[40] The food is donated by members of the public and by food retailers. Many critics of food banks argue that a charity offering food parcels is the wrong approach to tackle hunger, because it does nothing to redress the underlying causes that drive someone to need free food, such as low-paid and insecure jobs or changes to the welfare system.

Ede tells me he is very aware of the power imbalance between the users of food banks and the people doling out the food. "The people coming here are in a position of low power and the people giving out the food are in a position of high power." Ede says that the food bank volunteers do everything they can to make the people arriving to ask for food—many of them children—feel welcome and respected. But Ede tells me he often pictures how dreadful the first journey to the food bank must be.

As a food bank user, you are reduced to eating food that other people have chosen for you, which may or may not be what you wanted to eat. It's like being in a state of wartime rationing again, complete with the canned meat. If someone really doesn't like something in the box they have been given, they can take it out and choose something from the "extras table" instead. The extras table is full of what Ede calls "weird stuff," like jars of paté, unclassifiable canned fish, and olives and curious pickles that someone may have dug out from the back of a cupboard at home. In our age of consumerism, it seems harsh that the limits of someone's food choices should be reduced to rummaging on the extras table. Given the level of need, I am glad that food banks exist, but I can't help feeling we may look back on this phase in our food history with horror. How can a single city contain people so affluent they can have any food from any cuisine biked 'round on a whim and others so poor they are supposed to be grateful for a free can of tomatoes?

The food bank is a sticking plaster that cannot heal the gaping wound of hunger in the modern developed world. Food banks give people calories when what they really need is money and a decent place to live. There have always been those who went to sleep with a stab of emptiness in the stomach. The difference today is that so much hunger is happening in cities teeming with food. Hunger today affects people in full-time work as well as those without jobs. Hunger is never simply about a simple lack of food. It is also about a lack of social resources and about families who cannot afford to heat a house and pay the bills while earning only the minimum wage.

The food bank and the food pantry are the polar opposite of online grocery shopping. In a world where shopping has become a private activity governed by near-infinite choice, users of food banks are forced to collect their food in a public space where almost all the decisions and control have been taken away from them. In a food culture fixated with freshness and new flavors, the users of food banks are often reduced to canned food and dull seasonings, with many donations consisting of exactly the kind of ultra-processed food that most of us already eat too much of. Many American users of food banks are affected by type 2 diabetes and obesity, so to be given Hostess cupcakes and sugary sodas is not the most helpful move.

Over the past couple of years, there has been a drive at many food banks to look more closely at the health impact of the food that's being handed out. In 2016, the Capital Area Food Bank in Washington, DC, announced it would no longer accept donations of sugary foods such as sodas or candy.[41] Feeding America, the leading network of food pantries across the United States, has set a goal of 70 percent of the items distributed being "foods to encourage" such as fruits and vegetables and whole grains.

Food banks are not the only way to address food poverty. Other ways include the social supermarket (low-cost community-run food shops) and the food hub. With the food hub, the process of acquiring food becomes sociable again. The idea of a food hub is to create nonprofit warehouses and distribution centers that deliver produce grown by local farmers to local residents at low cost by cutting out the middleman, the commercial food supplier. Food hubs offer a tiny glimpse of what a shop might look like if it was designed neither to entertain its customers nor to extract maximum profit from them but simply to feed them. It makes you see how rare that sense of being nourished has become in the rest of our lives. For all our food choices, many of us act as if we are half-starved for the experience of being fed, as a parent feeds a child.

In 2016, New York City was given $20 million for a new Greenmarket Regional Hub, to deliver farmer's market produce to community programs and institutions.[42] Some of the beneficiaries of this food hub have been senior citizens who pay $1.50 to eat a lunch featuring organic vegetables at the Lenox Hill Neighborhood House, a senior citizens' center. Thanks to the New York City food hub, Antonio Perez, a retired hotel worker, tasted zucchini for the first time at the age of seventy-one. "It's very satisfying," he told a journalist for the *New York Times*.[43] The value of such a meal goes beyond the canceling out of hunger. It is about treating people with respect and giving them the kind of security and joy that any person might feel when they eat a meal out.

The whole way we talk about the acquisition of food in modern life has gone askew. We talk far too much about food as an amusing leisure activity and far too little about food in terms of basic human needs. To have access to regular meals of a decent quality, preferably eaten in company, is not some irrelevant optional extra—like deciding whether to have guacamole on your burrito. Access to food of a decent quality is something that every human deserves. For millions, there has never been such an exciting or abundant time as today to live and eat, but for all the plenty, we haven't yet figured out how to let everyone join the party. Eating a decent dinner may not be a duty anymore, but that doesn't mean that it shouldn't be a right.

CHAPTER SEVEN

EATING BY THE RULES

IN 1903, THERE WERE JUST EIGHT COLORS OF CRAYOLA crayon: black, brown, yellow, red, orange, indigo, violet, and green. Eight was the limit of a child's color choices, and selecting the colors needed for a picture was a simple task. If you wanted to draw the sun, you reached for the yellow. Easy. The downside was that with such a small box of colors, many things that a child might wish to draw would have been impossible to depict. Sky could only be the indigo of evening or the black of night, and the only way to do a cloud was to leave the page blank.

In 1935, something thrilling happened. The number of Crayola crayons doubled to 16, with the addition of such newfangled luxuries as pink and white. A world of color possibilities opened up, but it didn't stop there. In 1949, after the war, the number of colors vividly expanded to 48, with many shades of blues and peaches and purples. By 2010, the number of Crayola crayons was 120 and still growing. At the time of writing, the biggest Crayola box contains no fewer than 152 colors, including colors unasked for in the past, such as Neon Carrot, Razzle Dazzle Rose, and Timberwolf. From all this bounty, how does a toddler now decide which yellow is needed to draw the sun?

What has happened to crayons has happened to food, only more so. Part of the paradox of modern food is that we are eating in a world of choice that is impossibly and inhumanly large. Choice does not always set us free. Sometimes, it leaves us paralyzed with indecision—or so I find whenever I am faced with a restaurant menu that is too long and I start to feel that whatever I opt for, I will regret five minutes later when I see a more delicious option sailing off to someone else's table. I hanker for someone stronger and wiser to choose for me.

In the supermarket, I feel a different kind of indecision. The part of my brain that is still a compulsive eater resurfaces, and I sometimes worry I might eat everything in the shop. To navigate a supermarket and find the things that you want or need is to make a series of rejections. "Not this, not this, not this, not this . . . but *this*." It is cognitively exhausting to refuse so many items before you find the one that you actually want.

As we've seen, the average Western supermarket now contains nearly forty thousand SKUs (stock-keeping units), a level of choice that is both unasked for and unsettling. Back in 1930, the first supermarket contained just two hundred items. Such is the abundance of the nutrition transition. *And they never went hungry again.* For consumers if not necessarily for retailers, the ideal number of foods to choose from would be somewhere between two hundred and forty thousand, but where does the line fall, and who will draw it for us? Psychologist Barry Schwartz coined the phrase "the paradox of choice" to describe the fact that having too many options tends to make us less happy, not more. Studies suggest that when consumers are given a small selection of jam to choose from, they end up more contented with the flavor of jam they pick than when they are offered different types of jam from a larger array.[1]

"A child who has had a box of 48 crayons is never again satisfied with only 8," observed writer Angela Palmer in a 2017 essay on how our sense of what we "need" has escalated crazily in the modern world. "Or maybe," she adds, "they don't know how to ask for less in a society of more."[2]

One of the most surprising changes in our modern food culture is how many of us *are* suddenly starting to ask for less: to demand that some of the colors available to us are put back in the box. One sign of this is the rising popularity of no-frills grocery stores such as Lidl and Aldi in Europe and Trader Joe's (which is owned by Aldi) in the United States. The atmosphere of an Aldi store is very different from Trader Joe's. Aldi promotes itself on saving money whereas Trader Joe's specializes in organic and health foods such as "Take a Hike Trek Mix" and sunflower seed butter. But what these shops have in common is that they carry a mere fraction of the SKUs of the mainstream supermarkets. Aldi and Lidl stock between 1,400 and 3,500 lines, while a typical Trader Joe's carries 4,000 items. Mostly these are store-brand goods, so you don't have to waste time desperately weighing the benefits of one brand of canned plum tomatoes against another. As one enthusiastic marketing guru wrote in a blog about Trader Joe's, "Here you will find no redundant and confusing options to choose from!"[3]

Meanwhile, many consumers have started to find their own ways to close down some of the options of modern food. Every restaurateur I speak to says there has been a colossal rise over the past couple of years in the number of guests demanding foods tailored to a restrictive diet. A caterer at a British university told me that until a few years ago, when overseeing a black-tie dinner for two hundred people, he might expect to cater for no more than half a dozen special requests. Now, he's found that the number of people asking for particular foods has risen to around 50 percent of the total. What's more, many of the demands have become complex and specific. Diners have started making requests such as "vegan plus shellfish" or "only serve me meat if it's organic."

Special diets are no longer the preserve of those suffering from life-threatening food allergies such as celiac disease and peanut allergies—although these are also on the rise in the general population, for reasons that are not entirely clear. Above and beyond the epidemic of officially recognized allergies and intolerances (such as lactose intolerance), there is a second and much larger wave of people carving their way through the forest of modern choice with restrictive diets of their own making, whether it's cutting out wheat or carbs or milk or some customized combination. From 2009 to 2015, worldwide sales of nondairy milk alternatives such as almond milk and oat milk nearly doubled, and by 2016, they were worth $21 billion.[4]

I'm still a cow's milk person, myself, but I know ever more people who choose almond milk instead. Some prefer the taste, some feel it suits their digestion, and some want to opt out of the mainstream dairy industry. But in a curious way, the new vogue for plant milks is also a return to older times, when the rhythm of life included fasting as well as feasting. Almond milk is, in fact, nothing new. In late medieval Europe, creamy nut milks were what cooks used on fast days, when all animal products, including milk and eggs, were off the table. As well as almond milk, medieval cooks made almond butter and even almond cheese.

One of the most popular ways to experiment with eating right now is intermittent fasting—short periods without any food. Some use it for weight loss; others because they believe it gives them added energy or mental focus. I have a friend, not a faddish person, who eats not a bite from evening on Sunday to evening on Monday every week and says she feels much the better for it. Another version of modern fasting is the keto diet, which involves consuming almost no carbohydrates but large amounts of fats such as coconut oil, nuts, and butter, as well as meat. The result is to induce in the body a fat-burning state known as ketosis. Some studies have suggested that ketogenic diets can be beneficial for individuals with type 2 diabetes. Critics of the keto way of life say—accurately enough—that it is a deeply restrictive way of eating. But for those who swear by these diets, maybe the restriction is part of the appeal. Eating without limits can seem like a scary thing in today's world.[5]

From 2015 to 2016, Google data shows, there was a big leap in the number of internet search queries for dietary restrictions of many kinds. Popular searches covered the gamut from religious food restrictions to food intolerances. People searched, among other things, for "halal meat," "lactose-free milk," "best weight loss shakes," "vegan mac and cheese," and "is quinoa gluten-free?"[6]

In a world of forty thousand choices, the old advice of "everything in moderation" no longer cuts it. The signs are that many people have understandably had enough of this free-for-all of supersizing and hidden sugars, of type 2 diabetes and food waste. In the past five years, millions of eaters have rejected huge swaths of mainstream food and created their own rules to eat by. Such reactions offer a sliver of hope that eating—for some populations, anyway—is finally moving in a healthier direction, with a new thoughtfulness about food and a return to vegetables. On the other hand, some of the new diet rules we have invented for ourselves are as extreme and unbalanced as the food system they seek to replace.

BLEEDING BEETROOT

Vegetarianism—including veganism—is one of the biggest trends in eating right now. Out of a clear blue sky, people have started eating "vegetable main courses" such as cauliflower steaks or whole salt-roasted celeriac cooked to savory softness in an oven and carved up like a rib of beef.

There are still many more meat eaters than vegetarians in the affluent cities of the world, but vegetarian diets are growing as fast as cress, particularly among the young and health conscious. If stage four of the nutrition transition entails a huge rise in meat eating, the next stage of affluence—this is the hope, at least—would be to cut back again and return to plants. Between 1994 and 2011, the number of vegetarians in the United States more or less doubled, to seven million people. Still more striking is the fact that a third of American vegetarians are now vegans, who eschew not just meat but all animal products, including cow's milk and honey. In the United Kingdom, the number of self-proclaimed vegans has increased by 350 percent since 2006, from 150,000 to 542,000 (as of 2017). Suddenly, our shops are full of plant-based products that we never imagined in the past, such as coconut yogurt and even soy "calamari" designed to emulate the texture of fried squid.[7]

The vegetarians and vegans I knew as a student in the 1990s were resigned to eating things that they did not particularly enjoy for the sake of their ethics. They made do with dried-up nut roasts and drank coffee black because it was a rare café that would provide soy milk. Restaurant menus usually offered just one vegetarian option, dull and underseasoned and normally involving cheese—and sometimes none at all, in which case my vegan friends would order a couple of side dishes. Veganism in the 1990s was not necessarily even healthy because, in reality, it often meant dining on french fries and a side of olives.

New-wave veganism is completely different. When you speak to chefs now, it is clear that vegetables are an object of desire, as never before. Devising every meal around a big meaty centerpiece has started to look like an unimaginative, unsustainable, and wasteful way to eat.

In March 2018, I meet Alex Rushmer, an idealistic Cambridge-based chef in his thirties who is in the process of setting up a "vegetable-centric" restaurant. We are sitting in the kind of hipster café where you can get a brunch of cornbread with pickled chili, shredded carrots, smashed avocado, and burned lime. "Attitudes to vegetables are changing, and I think it's changing fast," Rushmer remarks. Though not fully vegetarian himself, Rushmer sees cooking with mostly plants as the answer to a series of misgivings he has long had about modern food, ranging from the inefficiencies of producing meat to the unhealthiness of the standard Western diet. Cost is another reason why Rushmer does not cook much meat, especially when he is cooking for himself and his wife in a nonprofessional capacity. "Our default meal at home is some kind of vegan lentil dal," he remarks, because to eat the kind of higher-welfare meat that he espouses costs so much. "And dal is *so* good."

I am increasingly struck by the idea that eating less meat is one of the easiest ways to winnow down the endless choices that accumulate in a standard supermarket. As the anthropologist Richard Wrangham has written, meat equates to joy in societies where food is scarce, but in affluent societies, eating less meat does not have to be a hardship.[8]

To me, eating more vegetarian meals—but not exclusively so—feels like a pragmatic path through the jungle of modern food options. It calms down my overloaded brain and frees me up to think more carefully about flavor and texture and what's in season right now. When I limit myself to mostly vegetarian foods, I am no longer quite so insistently plagued by the option paralysis about whether what I am buying will be healthy enough or sustainable enough or contain enough vegetables. I still worry about whether my picky and meat-loving youngest child will eat what I cook, but that's another matter.

"Only buy the best meat you can afford, grass-fed for preference," say a host of experts on ethical eating. But they do not explain where we are meant to find this grass-fed ethical meat as we nip into the nearest shop after a long workday looking for something quick for supper. It is true that there is a world of difference—in flavor, in nutrition, and in animal husbandry—between meat from an animal that has been slowly raised, well fed, and given freedom of movement and meat from animals fattened up on grain in confined and dark spaces. Offered the chance, spending more money on higher-quality meat is clearly a better way to consume meat—for humans, for the land, and for animals. Like most of our food supply, modern meat production is crying out for reform. But something rankles about describing this as an ethical form of meat eating when it is so far out of the reach of the budgets—including the time budgets—of most of the world's consumers.

For me, the best compromise has been to make meat a smaller element than it used to be in my family's eating without eliminating it altogether. This is becoming much easier than it once was because vegetarian recipes have become so exciting and flavorsome over the past few years. There is no deprivation in a meal of herb-green falafel with spicy hummus, warm flatbreads, juicy black olives, and pickled radishes. And then, at the weekend, I'll still make a chicken pie or a lamb curry. Some call this way of eating "flexitarian" or "reducetarian." To me, it is a way to redress the balance of our consumption a little.

Vegetarian food is undergoing a series of transformations, one being that so much of it is now eaten by nonvegetarians like me. For those immune to the charm of a lentil, there has been a striking rise in meat substitutes that taste uncannily like real meat—or least uncannily like real processed meat. In 2015, I was at a funfair in Philadelphia with my vegetarian sister and her three kids. We bought a range of burgers, some made from meat, some not. My sister spat out the first mouthful of her veggie burger because the texture was so meaty, she was convinced I'd given her the wrong one.

In certain cities, such as Berlin, there are now vegan butcher shops where you can purchase a gallimaufry of fake meats, from vegan charcuterie to "no beef strips" and "no chicken shawarma." The best estimates say it will be a few years before in-vitro meat (also known as lab meat) is in our shops: cruelty-free meat cultured from animal cells. But in the meantime, plant-based burgers have become so uncannily accurate in their imitation of meat that it's hard to tell the difference. It wasn't long ago that beetroot was an object of horror in our culture, whereas now young people are actively seeking out vegan burgers that bleed dark purple beet juice. The Impossible Burger—which has the backing of Bill Gates, among others—echoes the chewy texture and speckles of fat in a beef burger using coconut fat mixed with potato protein. An Impossible Burger will sear and even "bleed" like real ground beef, although the jury is still out on whether it will convince the world's billions of meat eaters to step away from the beef. Reviews of the Impossible Burger's taste have been largely positive. In 2016, business journalist Linette Lopez remarked that the Impossible Burger was very good, although to her it didn't "necessarily taste like a cow's meat . . . It's more like a reminder that cows exist somewhere out there."[9]

There are countless reasons we might want to eat less meat or no meat. Some do it for health reasons, others out of concern for the shocking way that animals are treated in mass-market meat production. Since 2015, a series of polemical provegan documentaries on the streaming service Netflix have driven many young people to eliminate animal foods from their diet. I met a man in his twenties recently who said he and his girlfriend had gone vegan overnight after watching a film about how animals are treated in the meat and dairy industries. One day, he was a bacon lover who adored meat so much he sometimes ate ground meat raw from the pack. The next, he would not eat so much as a strand of egg noodles or a spoonful of honey. He said there wasn't a day when he didn't miss bacon, and as a keen runner, he got very hungry unless he carried with him a supply of jam or hummus sandwiches. But knowing what he now knew about meat, he chose to regard it as something that was no longer food for him. Judging from the number of vegan options springing up in high street cafés and all-vegan menus in restaurants, he isn't alone.

In many ways, the new veganism looks like a hint that the human diet is moving in the right direction. But around the world, I have also met a number of people who observe that modern eating is becoming angrily polarized. The nutrition world has split into vegans on the one hand and on the other, extreme low-carb meat gurus such as self-help author Jordan Peterson, who claims he cured his depression by eating nothing but beef. Between these two polarities, the middle ground of eating can get lost. In Mumbai, Vikram Doctor told me that in coastal regions of India such as Kerala, there used to be a number of tasty dishes that were neither vegetarian nor fully meaty but somewhere in between: mostly vegetables seasoned with small amounts of seafood for added flavor and protein. Vikram has noticed that the tradition of these dishes is dying out, because they find favor neither with carnivores nor with vegans.

Many people have started to talk about food in an all-or-nothing way. Meat versus vegetables. Carbs versus fat. Superfood versus junk. "It's the guilt I don't like," said a friend of my teenage daughter. It wasn't enough to be vegetarian anymore, she explained. At her school, she told me, anyone who eats a slice of cheese can feel judged for not being ethical or healthy enough. This is the problem with inventing new food rules: once you start, it's hard to know where to stop. For some health-conscious modern eaters, rethinking meat is only the beginning on a journey to eliminate everything from our diets that is supposedly not pure enough.

THE HEALTHIEST OF ALL POSSIBLE DIETS

At first, clean eating sounded modest and even homespun: rather than counting calories, you would eat as many nutritious home-cooked substances as possible. This was about taking all the toxic crayons out of the box and leaving only the good ones. But it quickly became clear that in many cases, clean eating was more than a diet; it was a belief system that propagated the idea that the way most people eat is not simply fattening but impure. Seemingly out of nowhere, a whole universe of coconut oil, dubious promises, and spiralized zucchini has emerged. "I long for the days when clean eating meant not getting too much down your front," the novelist Susie Boyt joked in 2017.

The rise of eating for wellness is a sign that millions of eaters have become so unmoored that we will put our faith in any master who promises us that we, too, can become pure and good and glowing with health, if only we can remove enough foods from our diets.

It is very hard to pinpoint the exact moment when clean eating started, because it is not so much a single diet as a portmanteau term that borrowed ideas from numerous preexisting diets: a bit of Paleo here, some Atkins there, with a few remnants of 1960s macrobiotics thrown in for good measure. Sometime in the early 2000s, two distinct but interrelated versions of clean eating became popular in the United States—one based on the creed of real food and the other on the idea of detox. Once the concept of cleanliness had entered the realm of eating, it was only a matter of time before the basic idea spread contagiously across Instagram, where fans of #eatclean could share and parade their artfully photographed green juices in Mason jars and rainbow salad bowls.

The first and more moderate version of clean food originated in 2006 when Tosca Reno, a Canadian fitness model, published a book called *The Eat-Clean Diet*. In it, Reno described how she lost seventy-five pounds and transformed her health by avoiding all overrefined and processed foods, particularly white flour and sugar. A typical Reno "eat-clean" meal might be stir-fried chicken and vegetables over brown rice, or almond-date biscotti with a cup of tea. In many ways *The Eat-Clean Diet* was like a hundred other diet books that came before, advising plenty of vegetables and modestly portioned home-cooked meals. The difference was that she did not call it a diet at all but a holistic way of living.

Meanwhile, a second version of clean eating was spearheaded by Alejandro Junger, a former cardiologist from Uruguay. *Clean: The Revolutionary Program to Restore the Body's Natural Ability to Heal Itself* was published in 2009 after Junger's "clean" detox system had been praised by actress Gwyneth Paltrow on her Goop website. Junger's clean system was far stricter than Reno's, requiring, for a few weeks, a radical elimination diet based on liquid meals and a total exclusion of caffeine, alcohol, dairy and eggs, sugar, all vegetables in the nightshade family (tomatoes, eggplants, and so on), red meat (which according to Junger creates an acidic "inner environment"), and other foods. After the detox period, Junger advised very cautiously reintroducing "toxic triggers" such as wheat ("a classic trigger of allergic responses") and dairy ("an acid-forming food").

To read Junger's *Clean* is to feel that everything edible in our world is potentially toxic. "Who is the candidate for using this program?" Junger asks. "Everyone who lives a modern life, eats a modern diet and inhabits the modern world."

We are living in an environment where ordinary food—which should be reliable and sustaining—has come to feel noxious in ways that we can't fully identify. In prosperous countries, large numbers of people, whether they want to lose weight or not, have become understandably scared of the modern food supply and what it is doing to our bodies. When normal diets start to sicken people, it is unsurprising that many of us should seek other ways of eating to keep ourselves safe from harm. Our collective anxiety around diet has been exacerbated by a general impression that mainstream scientific advice on diet—inflated by newspaper headlines—cannot be trusted. First these so-called experts tell us to avoid fat, then sugar, and all the while people get less and less healthy. What will these experts say next, and why should we believe them?

Into this atmosphere of anxiety and confusion stepped a series of wellness gurus offering messages of wonderful simplicity and reassurance: eat this way and I will make you fresh and healthy again. These gurus were able to find receptive readers and followers in part because of the failures of conventional medicine to address the question of food. Over the past fifty years, mainstream healthcare in the West has been inexplicably blind to the role that diet plays in preventing and alleviating ill health. When it started, #eatclean spoke to growing numbers of people who felt that their existing way of eating was causing them niggles, from weight gain to headaches to stress, and that conventional medicine was unable to help. In the absence of nutrition guidance from doctors, it was a natural step for individuals to start experimenting with cutting out this or that food, from dairy to gluten. The gurus of wellness urged us on in such restrictions. Amelia Freer, in her 2015 book *Eat. Nourish. Glow*, admits that "we can't prove that dairy is the cause" of ailments ranging from irritable bowel syndrome to joint pain but concludes that it's "surely worth" cutting out dairy anyway, just to be safe. In another context, Freer writes, "I'm told it takes 17 years for scientific knowledge to filter down" to become general knowledge while advising that we should all avoid gluten in the meantime as a blanket precaution.

On the basis of such pseudoscience, millions of us have started to look with terror on a range of basic and nourishing foods, including pasta and noodles, which are sometimes referred to in the clean eating movement as "beige carbs." Like fat in the 1980s, gluten has come to be viewed by some as a pollutant, something that could contaminate a whole plate of food with a single speck. Only around 1 percent of the population actually suffers from celiac disease, an autoimmune disease in which the gut lining is damaged by gluten, the protein found in wheat, rye, and barley. A further small segment of the population may be suffering from a much milder—and still controversial—condition known as nonceliac gluten sensitivity, whose symptoms include brain fog, stomach pain, and bloating. But these two conditions together cannot begin to explain why one hundred million Americans, or a third of the population, say (according to industry data) that they are now actively avoiding gluten. I have celiac friends who claim that #eatclean has made it far easier for them to buy ingredients that they once had to go to specialist health food stores to find. But it is sad that so many other people should have forced themselves to abandon once-beloved foods and buy expensive gluten-free replacements for no good reason, in the name of wellness.

For as long as people have eaten food, there have been diets and quack cures, but previously, these existed, like conspiracy theories, on the fringes of a food culture. The clean eating of recent years is different because it has established itself as a challenge to mainstream ways of eating, and its wild popularity over the past few years has enabled it to move far beyond the fringes. Powered by social media, it has been more absolutist in its claims and more popular in its reach than any previous school of nutrition advice, outside the systems of dietary taboos created by religions (including Jain vegetarianism, a form of eating by the rules that is enjoying a resurgence in popularity).

ALMOST AS SOON AS IT BECAME UBIQUITOUS, THE CLEAN eating movement sparked a backlash, with critics arguing that the gurus were promoting a judgmental form of body fascism that encouraged their followers to view food as dirty. As the negative press for clean eating intensified in 2016–2017, many of the early goddesses of #eatclean tried to rebrand—declaring they no longer use the word *clean* to describe the recipes that have sold them millions of books. Yet however much the concept of clean eating has been logically refuted and publicly reviled, the phenomenon itself shows few signs of dying. Even if you have never knowingly tried to eat clean, it's impossible to avoid the trend altogether because it has changed the foods available to all of us and the way they are spoken of.

The influence of clean eating is present in turmeric lattes and "wellness bowls," in "guilt-free" snacks and "grain-free" salads. Some of these new items make a welcome change from the unwholesome foods of stage four. It makes me happy to be able to pick up a salad of roasted beetroot and feta from the kind of high street café that would once have sold only dull stodgy sandwiches.

The less appealing aspect of clean eating is a tone of moralism that threatens to undermine the pleasure of everyday meals. A young student friend of mine told me that the experience of eating out had been ruined by the fact that she can't order pasta in front of some of her friends without them preaching to her about the evils of carbs.

It's increasingly clear that an obsessive focus on eating only for wellness, for all its good intentions, can cause real harm. Renee McGregor is a British dietitian who works with both athletes and people who have eating disorders. From 2016 onward, McGregor has noticed, she tells me, that "every single client with an eating disorder who walks into my clinic doors is either following or wants to follow a clean way of eating." This is not to say that clean eating necessarily causes eating disorders, mental illnesses that have deep and complex causes, some of them genetic. But for the vulnerable people McGregor sees in her clinic, clean eating provides a set of seductive rules that, once lodged in their minds, make it much harder for them to recover. McGregor's own definition of healthy eating is "unrestrained and uncomplicated eating." By this she doesn't mean binging but eating from all the major food groups without guilt or fear.

A new form of eating disorder has taken root that has been given the name "orthorexia," from the Greek *ortho*, meaning "correct," and *orexis*, meaning "appetite." Orthorexia—about which McGregor has written a book—is an obsession with eating only food that is pure. Though similar to anorexia in many ways, the primary obsession for those suffering from orthorexia is not with losing weight but with restricting their eating to foods that have nothing "bad" in them. First, someone might cut out sugar, then meat and dairy, then bread, then all carbs, then saturated fats, then most other fats, then fruits (which are nothing but sugar), and before you know it, all you are left with is vegetables (excluding dangerous nightshades such as red peppers) and perhaps some nuts. McGregor writes that this "endless obsessive pursuit of rainbow's end" is "exhausting and depleting and unhealthy."[10]

A couple of years ago, McGregor saw a young client in her twenties who worked in banking. This woman had become worried she was drinking too much alcohol and stopped going out to socialize with work colleagues. By the time she arrived at McGregor's clinic, she was eating nothing but a small piece of fish or chicken and a plate of spiralized vegetables twice a day. She had come to see McGregor because her cholesterol levels were surprisingly high, and she wanted to know what else she could cut out of her diet to bring her cholesterol down. McGregor explained that it was her "healthy" diet that was giving her high cholesterol. This woman's half-starved body was no longer making enough estrogen, and this in turn disrupted her body's regulation of cholesterol. But she still believed that what she needed was to find one more thing to cut out.[11]

The true calamity of clean eating is not that it is entirely false. It is that it contains a kernel of truth. Underneath all the nutribabble talk of "glowing" and wellness, the gurus of clean eating are completely right to say that most modern eaters would benefit from consuming less refined sugar and processed meat and more vegetables and meals cooked from scratch. The problem is that it's near impossible to pick out the sensible bits of clean eating and ignore the rest. Whether the term *clean* is used or not, there is a new puritanism about food that has taken root very widely.

Once you start on a path of cutting out this unclean food and that one, you may start to feel that no food for sale in the shops can ever be good enough. No vegetable can be organic enough. The logical end point of this kind of diet absolutism is to send back the whole box of crayons and seek a substitute for food itself.

THE ENIGMA OF THE PROTEIN BAR

"He won't go out to restaurants with me," sighed a twenty-something woman I met in Seattle in 2016. She was despondent that her boyfriend refused to share her great love of food with her. In their relationship, there was no staring at each other across a candlelit table, no swapping spoonfuls of dessert, no grocery shopping for interesting new ingredients. In place of all this mutual joy, he simply ate vitamin-enriched high-protein "sports bars" whenever he was hungry, feeling no need to share. He was a gym fanatic and claimed that the bars gave him all the nutrients his body needed. "At least he's healthy, I guess," she said, not sounding very convinced. She looked sad and asked me if I thought he would ever change. She reminded me of a parent whose child won't try anything green.

When someone eats nothing but pizza and string cheese, we call them a picky eater. Yet when someone builds their life around modern meal replacements such as protein bars and "complete" liquid break-fasts, our time-obsessed culture may encourage them to believe they are doing something superior to the drab old business that the rest of us indulge in when we sit at a table with knives and forks and con-versation. In this screen-fixated age, such people are carving out more minutes and hours to spend on *useful* stuff like email and social media and ordering more possessions online.

When choosing how to feed ourselves, many are searching for something that will fill us up while containing minimal calories. We want it to help us with sport and with work. We'd like it to be totally portable and for it not to require that we sit down to eat it. We want it to have nothing "bad" in it, however that might be defined. We expect it to deliver protein to our bodies while tasting as sweet as candy. Sad to say, there are plenty of things we desire from food now that a traditional meal simply can't deliver. Enter the snack bar.

At a trade show in London in the winter of 2016, I started talk-ing to a man from the Netherlands who worked for a company that designs and creates bars for anyone in the food industry who wants to launch an edible bar, whether it's a sports bar, a protein bar, a bar for serious trekkers and hikers, a raw vegan bar, a functional vitamin-packed lunch substitute, an energizing pick-me-up for middle-aged women, or a weight-loss bar for young slimmers. The man handed me a thick brochure. "Over the years, we've . . . created bars people have only dreamed of," it said. The company can develop a bar that is light and crispy with puffed rice or dense and compacted with nuts and chocolate. The inside of the bar can be layered with fruit jelly, coated with solid yogurt, or swirled with sprinkles like a cake.

It's hardly possible to compute the number of rival snack bars in the world today or to count the number of different human needs they claim to satisfy. Back in the 1980s, edible bars existed on the fringes of our eating lives, the preserve of athletes and health food enthusiasts. At that time, crunchy oat and almond cereal bars were sold six to the pack as a snack for those with wholesome tastes. There were also dense, chewy energy bars sold singly for serious cyclists and runners; PowerBar was the leading example. For such a bar to be eaten on a whim by a nonathlete would have seemed unimaginable.

It was only in the 1990s that the "nutrition bar" went mainstream. It started with the Clif Bar, launched in 1992, which aspired to unite the nutrition of an energy bar with the texture of a biscuit. In 1992, Clif Bar's annual sales were valued at $700,000. By 2002, they were worth $106 million, and Clif had been joined by countless rival bars. The majority of Americans are now regular snack bar consumers, with six out of ten adults purchasing snack bars during a three-month period in 2017, according to data from Mintel.[12]

The idea of the snack bar was to simplify our food choices, but in reality, it has added yet another set of alternatives. We are talking about a whole new box of crayons. In 2005, there were 750 different bars on the market in the United States. By August 2017, the number of snack bar varieties for sale in the United States was closer to 4,500—more than the total number of items available at a Trader Joe's or Aldi.[13]

This growth has gone along with minute levels of differentiation. It's no longer special enough for a bar to claim that it is all natural. To stand out in such a crowded market, a snack bar needs to go niche. There are keto bars and bars for people on a gluten-free diet, bars for kids and bars for pregnant women.[14] Bars proved to be a gift to food companies wanting to tap into markets for different diets, from sugar-free to plant-based. What diet will bar makers profit from next? "Diabetes is a totally undeveloped market," suggested a market report on sports nutrition from 2017.

It's easy to see why manufacturers would want to sell us so many bars. But why, I asked the snack bar creator from the Netherlands, do we as consumers choose to buy so many of them? It's not as if these bars are cheap. Bought singly from the gym or a convenience store, they can easily cost as much as two dollars each. "A lot of people now have disordered eating," was his frank reply. "They are cutting out this ingredient and that ingredient." He also attributed the rise of bars to the increasing numbers of people who do serious exercise as a hobby, who want extra protein to help them achieve their goals.

The protein bar is one of the enigmas of our times. Why is it called a protein bar and not a sugar bar? I picked up a selection of protein bars from my nearest supermarket. Some were studded with peanuts and some were in the form of a coconut flapjack. Yet no matter what form the bar took, they all contained more carbohydrate than they did protein, with much of it coming from sugars such as glucose syrup or honey. Every last protein bar I tasted had the sweetness of a biscuit (albeit a very expensive and slightly disappointing biscuit).

When talking of food, the word *protein* would normally suggest something savory and hearty—an egg, a slice of roast pork, a cube of salt-and-pepper tofu, a bowl of spicy black beans—so it's odd that the protein in a protein bar does not come in savory form. "Many people have tried to make a savory protein bar," my informant tells me. "They launched it in Italy and Austria. People keep asking for it, but when it's on the shelf, they don't buy it. We don't really know why, but the typical protein bar consumer wants it to taste like a brownie, a dessert."

Like so many other things we now consume, the protein bar is a paradox. It is a license to eat candy and call it a virtuous main course. In our sugar-laden environment, it's yet another sweet food, yet somehow one that convinces us that it is different: nothing so common as a biscuit or a scone. Where real food is infinitely varied in flavor, the meal-replacement bar almost always tastes of either chocolate or nuts, and often both at once.

And yet this pantheon of bars still has some residual connection with real food. They may be sweet, but they do, at least, offer something to chew on. Gary Erikson, creator of the Clif bar, has written that he grew up eating traditional food cooked by his Greek mother. He espouses the values of "slow food" and has sometimes struggled

with the contradictions of making an energy bar designed for convenience and portability.[15] Yet it's Erikson's hope, he has written, that
people will at least "slow down a bit as they eat our bar" and savor the
sensation in the mouth.

By contrast, little savoring goes on when you drink a liquid meal
replacement such as Soylent or Huel. Nor should it, according to the
people who make them. These new liquid foods are designed to liberate consumers not just from choice but from the distractions of flavor
and texture.

BEYOND FOOD

In the past, food was always a given in human life. Some ate one food,
and some preferred another. Some hated turnips, and some despised
licorice. But there was no question of turning your back on food itself—
unless you were suffering from an eating disorder such as anorexia
nervosa. Something very strange and new is happening in our society,
where over the past decade, a sizeable minority of people has come to see
not eating as a better option than eating. "I haven't eaten a bite of food
in thirty days, and it's changed my life," announced the entrepreneur
Rob Rhinehart in a 2013 blog on his invention of Soylent, the world's
most popular meal-replacement drink. Millions are now following in
Rhinehart's footsteps, whether by replacing food with products such as
Soylent or simply by cutting food out for days at a time.

In Silicon Valley, it's become popular to engage in "biohacking"
through periods of fast that put the body in the metabolic state known
as ketosis. Geoffrey Woo, the CEO of a biohacking company who led
more than a hundred people in a seven-day fast in early 2017, has
said that he sees the ketones generated by fasting as a kind of "super-
fuel for the brain." After two or three days of not eating, Woo finds
that the hunger "tapers off," and he develops feelings of mental clar-
ity. Experts on eating disorders, however, have warned that extended

fasting can be dangerous and elide into other forms of food avoidance such as anorexia.[16]

To me, the thought of going days without food remains dystopian. Waking first thing, I am often in a state of mild panic: Why didn't the alarm go off sooner, how will I make inroads on my to-do list, what do the children need for school today, and why didn't I get it ready the night before? What calms me is the prospect of coffee and toast followed by yogurt and perhaps a pear: that warm, comforting smell filling the kitchen and the reassuring music of the coffee grinder and the toaster. Food makes the panic dissipate.

For those who consume meal replacements, on the other hand, food itself has clearly become the cause of panic. Maybe the protein shake and the sports bar are logical reactions to an age that can't stop talking about food, with Instagram feeds devoted to cake and restaurant meals photographed as fondly as if they were babies. For some people, the complexity of feeding ourselves—the ethics of green beans, the healthiness or otherwise of butter, the myriad choices of sandwich fillings for a quick working lunch, the sheer expense—is just too much. With a flask of reconstituted meal-replacement drink, you can opt out of all this complexity, knowing that you have covered your nutritional bases. It's a life hack. All that you need do is swallow and get on with your day.

Meal replacements are surprisingly widely used among the population at large, and their use is growing at a prodigious rate. Start-ups the world over are making fortunes from selling sacks of various powders to turn into meal-replacement drinks.

Around a million people in the world tried a drinkable meal in 2016, of whom two hundred thousand were regular users, the vast majority of them men, for reasons that are not yet clear. My own theory about why women are more reluctant than men to swap meals for a bland drink is that women are—on average—much more genetically sensitive to aromas and flavors and therefore may feel more reluctant to give up solid food.[17]

When my grandmother lay dying, fifteen years ago, she could no longer chew her food easily and was reduced to drinking a series of meal-replacement powders that kept her alive for another day but gave her no joy. She would have been astonished to learn that a few years later, fit young people would be actively choosing to drink such products and calling them better than food. The US-based Soylent, launched in 2014, is the most famous, but there are now over a hundred competing brands of these powders worldwide, including Queal, Joylent, Mana, KetoSoy, and PowerChow.

Does anyone actually relish a meal replacement? For all the billions of dollars we collectively spend on them, I've never heard anyone say they *adore* a protein bar in the way that you might rhapsodize about sinking your teeth into a buttery head of corn on the cob or nibbling a rich slice of date-and-walnut cake, crumb by delectable crumb. The meal-replacement consumer selects these products less for how they taste than what we imagine they will *do* for us. Will it kill our hunger over four hours? Will it help our muscles recover after lifting weights? It's like pet food for humans.

The idea that it might be better to gulp our food as fuel rather than to savor it represents a profound shift in our eating habits, even if replacing all meals in this way is still a minority habit. For most of human history, the aspect of eating that really compelled us was what happened on the tongue and in the nose. Freedom from poverty meant being able to afford to eat with relish and enjoy varied sensations rather than being limited to the bland staple monotony of rice or bread. Meal replacements represent a return to monotony. I can't pretend to know what the gruel of our ancestors tasted like, but I imagine it to be something like the squeezable cold oats now widely sold in aluminum pouches in lieu of a solid breakfast.

The fact that we are turning to these items in such quantities—despite their relative lack of flavor and texture—reveals once again what an alarming thing food has become in the modern world for many people. The real question here isn't why millions now eat sports bars and drink meal replacements every day, but what we believe we can get from these products that we can't get from food.

THE OPPOSITE OF A CUCUMBER

If you want to make Julian Hearn angry, get him talking about cucumbers. Like other proponents of powdered meal-drinks, Hearn, the cofounder of Huel—a drink engineered from pea protein and brown rice in the United Kingdom—argues that we have "optimized" the wrong things when it comes to food, as he tells me one day in 2017. "People have been brainwashed into thinking that food is about taste and texture, not nutrition." As a cook, when I see a cucumber, I feel the opposite of rage. I think of mint leaves and feta cheese salads and cooling summer soups. A cucumber seems too mild a thing to be offended by: so juicy, pale, and pleasant. But when Hearn sees a cucumber, he sees only waste and inefficiency.

What riles Hearn about cucumbers, he tells me, is that it's a food that prioritizes texture over nutrients. Sure, it may be nice and crunchy to put in your mouth, but the resources that go into producing it cannot be justified, in his view. "It's grown in a hothouse in Spain, transported on a chilled train, and processed in the UK before it comes to the table." Any cucumber that does not meet the supermarket's strict requirements regarding shape will be wasted. Yet at the end of this labor- and resource-intensive process, what do you have? To Hearn, the cucumber is "basically water and only 3 percent nutrients." By contrast, Huel aims to put nutrients first and flavor second.

I can't tell you whether drinking daily meal replacements is actually good for your health or well-being, because these products are so new that few studies have been performed on the long-term outcomes of consuming them. Most of the science on meal replacements comes from studies that follow people drinking weight-loss shakes. One study on the effects of drinking a soy-based meal replacement twice a day found that only 5 out of 155 subjects experienced stomach problems or diarrhea, while the rest found it "well tolerated." To me, that is not a ringing endorsement, but I guess it depends what you are looking for from your food.

Huel boasts that it contains "all the proteins, carbs, and fats you need," plus twenty-six vitamins and minerals, while also being vegan. Along with pea protein and rice, it contains flaxseed, coconut, vanilla flavoring, and various other carefully chosen bits and pieces. This is something to "take," much as you would take a vitamin pill, rather than linger over, like a ripe peach.

Step on a bus or a metro in any major city between the hours of 6:00 a.m. and 4:00 p.m., and there will be at least one person clutching a shakable blender bottle full of some beige meal-drink or other. Often the drinker is wearing yoga clothes; other times he or she is staring at a laptop or a book, because the drink itself requires no attention.

Huel is designed to be the opposite of a cucumber. It is nutrition first and food afterward (although for those still hung up on pleasure, the Huel website also sells flavor sachets, including pineapple, coconut, or chocolate, plus seasonal one-offs such as a Christmas pudding flavor, designed for Huel-drinkers suffering wistful pangs of desire for real Christmas dinner). The name *Huel* is a fusion of "human" and "fuel," and it's Hearn's ambition to make us see that we have overrated the sensory joy that food gives us at the cost of the nutrients it contains. By the same token, Rob Rhinehart, the inventor of Soylent, the world's most successful powdered food, has argued that Soylent's relative lack of flavor is a sign of what an effective technology it is. "Water doesn't have a lot of taste or flavor and it's the world's most popular beverage," Rhinehart told *The New Yorker* in 2014, the year his product was launched.[18]

For Rhinehart, Soylent was originally a frugal answer to the fact that food had become a "burden" to him. At the time he invented the formula for Soylent, he was a twentysomething living in San Francisco working on a tech start-up, subsisting on an unhealthy diet of instant ramen, frozen quesadillas, and corn dogs, plus vitamins to improve his nutritional balance. When he first developed Soylent, Rhinehart discovered he could save more than $400 a month and feel healthier to boot by drinking a self-made powdered drink mix for all his meals: a kind of supercharged milkshake. He wrote a blog post entitled "How I Stopped Eating Food," extolling the efficiency of living without food.

Hearn's take on Huel is slightly different. He doesn't see it so much as an anti-food as the fulfilment of food's original promise, which was to be essential, not a distraction. Huel is about getting away from what Hearn sees as the inessentials of texture and flavor and back to nutrition.

Hearn's view genuinely seems to believe that a vegan powdered meal such as Huel can begin to offer an answer to all the problems of modern food, from obesity to food waste and the inhumane treatment of animals destined to become meat. But it will involve a trade-off with the world of spice and joy that we currently enjoy. "Today we are flooded with flavor and taste," he remarks, citing the huge proliferation of different cuisines in modern Britain, from Indian to Thai and Korean. "I mean, I never tasted a curry until I was twenty." Hearn complains that in today's world, we have too many food choices and that too many of those choices are governed by the logic of flavor rather than of simple nourishment. As I'm talking to Hearn, I suddenly realize that for all the talk of body hacking and macronutrients, there is something old-fashioned about Huel. I sense a wish in Hearn to return to the spartan eating of a postwar childhood when dinner was underseasoned meat and two veg and, as he puts it to me, there was "just so much less choice."

Food fulfils two needs in our lives—pleasure and fuel—which are usually inextricably intertwined. The meal-replacement drink offers an unprecedented opportunity to separate the functions of food out again and to discard everything about food that is recreational, or at least to save it for another occasion. It shuts down your options.

Most users of these complete powdered foods do not switch to them wholesale, choosing instead to use them for weekday breakfasts and/or lunches, keeping the fun and social aspect of food for evenings and weekends. Hearn has a wife and a young child, and on the weekends he enjoys leisurely meals with them. He admits that if you have access to a kitchen and know how to cook what he calls a "whole food meal" (preferably without any cucumber) then "that meal will be superior to Huel" and says that his advice would not be to switch to Huel "100 percent or even 90 percent" of the time. But he insists that it is still the ideal solution for meals when you are away from a kitchen. I decide to take him at his word and try out a week of Huel lunches.

You know that scene in *The Wizard of Oz* where Dorothy leaves the black and white of Kansas and is plunged into the magical Technicolor of Oz? For me, the first day of switching from a normal lunch to drinking Huel was like going from Oz back to Kansas. It felt like leaving a world of color and being plunged into gray.

Normally, I spend half my morning thinking about lunch and wondering how early I can justify stopping work. On my Huel days, by contrast, I found that lunch itself became the thing I procrastinated over. I was dreading opening my flask of Huel and feeling the cold liquid in my mouth. Eventually, hunger won out over disgust and I sat down to drink it. It smelled like cake batter. The texture was slimy and grainy at the same time. About halfway through my allotted portion, I felt so full I wasn't sure I could carry on, but I forced it down. A sudden melancholy came over me: no more eating until dinnertime.

Health-wise, I didn't feel very different on my Huel lunches, except for a mild queasiness, and I neither gained nor lost weight (although five days is not really long enough to judge). To my surprise, the drink did a great job of canceling out my hunger pangs until dinnertime. What it couldn't do was eliminate my craving for the flavors and textures of solid food. At the end of my five days, I returned, with gratitude and relief, to lunches of real food. The colors and flavors were almost shocking. I hadn't appreciated before quite how much my daily mood was boosted by my working lunch, even when it was a box of leftovers. Purely in the macro terms of proteins and carbohydrates, meal-replacement drinks may offer perfect nutrition, but the psychology of consuming them is another matter. Given the choice between a flavorsome and wholesome lunch and a powdered drink, I can't imagine choosing the latter.

But what if the choice were not between a meal replacement and something delicious but between one of these drinks and a lunch that is greasy, flavorless, and overpriced? In short, what if the other option was the average lunch foods available to most busy people on a budget today?

To contemplate the reasons someone would have for using meal replacements is to see how disappointing the experience of eating has become for many. I started to think about meal replacements differently when I came across a blog written by Dan Wang, a twenty-five-year-old writer and editor. When I first contacted Wang, he was drinking Soylent once a day, and he argued that the prejudice against these drinks is misplaced. Unlike many users, he did not claim that Soylent had changed his life or made him healthier. He did not pretend it was even particularly satisfying. Neither, he pointed out, is most of the food for sale in the average Western city: "I challenge the doubters to declare that every meal they have is a plate of nutritious deliciousness, prepared simply, and enjoyed in the company of friends."

Wang himself started drinking Soylent in 2015, not to replace all meals but those that would otherwise become "a hot dog in a cafeteria"—in other words, lunch. Soylent comes in two versions: "Version 1.3, which I'll call cake-mix Soylent; and Version 1.4, which I'll call burnt-sesame Soylent." Wang marginally preferred the cake-mix version, although he did sometimes worry that he would one day find it too disgusting to swallow. When he first heard about Soylent, he assumed it was a "massive joke," but before long, the mixing of the powder and water became normal for him. "The main thing was not having to think much about food," he told me when we spoke. "That was really good."

For Wang, Soylent made sense in modern American life, not because it was utopian and marvelous, but because the rest of the Western food supply was so dispiriting. "I spent the first seven years of my life in Kunming, Yunnan," Wang told me. He saw Kunming "as a food paradise: the broths, the rice noodles, the cold dishes, and especially the mushrooms, all delicious." When his parents left China and moved first to Toronto and then to Philadelphia, he found himself feeling "food homesick" every day. But at least he had the family food to look forward to in the evenings. The authentic dishes his parents cooked kept him connected with the joys of Kunming even in North America.

When Wang left home and found himself studying at the University of Rochester in New York State, none of this deliciousness was available to him. His lunch options—the only ones he could afford on a student income, anyway—were "microwave meals and greasy hamburgers." He could have bought a pricy salad, but he does not care much for salad. The university's dining halls served food that was both expensive and off-putting. Wang turned to Soylent not because he didn't like good food but because he liked it too much to suffer the bad kind. He still cooked tasty Chinese meals for himself in the evenings—big pots of greens and rice with slow-cooked pork ribs or a chicken broth. But during the day, he found carrying a thermos of Soylent to class to be a way to avoid wasting money on bad food. It appealed to his sense of frugality.

"Soylent isn't a replacement for eating," Wang told me over the phone one day in 2016. "It's a replacement for *not* eating." After graduating, Wang moved to San Francisco. Based on two brief visits, I personally think of this sunny West Coast city as one of the best places to eat in the world. San Francisco summons to mind incredible citrus fruits (blood limes, Meyer lemons), damp sourdough bread from Tartine bakery, and the roast chicken and currant salad at Zuni Café. All those treats come at a hefty price, however, and Wang assured me that on his income, in the part of the city where he was working, he couldn't afford to buy what he considered a good lunch. He was still drinking Soylent once or twice a day, but he emailed to say that "if cheap, nutritious, and delicious food were easily available, I would opt for it over Soylent every time."

Much of the criticism of meal replacements has centered on the premise that "food" is necessarily joyous, sociable, and health-giving. Dan Wang was riled, he told me, that so many critiques of Soylent came from people who themselves had "excellent access to food." He resented the implication that "everyone should be eating Alice Waters–style organic vegetables at every meal. Very few of us are able to realize that even once a day."

The last time I spoke to Dan, it was late 2016, and he had stopped drinking Soylent. A new and better-paid job had taken him to New York City, where he finally had access to the appetizing Chinese food he had hankered after for so long. His new office supplied—for free— good-quality savory cafeteria lunches with plenty of vegetables, and every morning, Wang went out for breakfast at one of two Chinese bakeries in the neighborhood where he lived. He especially liked a Hong Kong bakery, where he could start the day luxuriating over dim sum and congee and salted egg and pork. His whole breakfast cost just five dollars. He offered to take me there if I ever visited New York. "I feel like a much healthier person than I was before," he told me. "Is this way of eating more varied than Soylent? Yes it is." But he said he would definitely drink Soylent again, if he ever again found himself in a situation where the only food available to him was the standard American diet.

It's a depressing comment on our bewildering food supply that not-food can now seem like a better option than food to thoughtful people like Dan Wang. Few people would thirst for Soylent in a world where wholesome and tasty food was easier to come by. "My definition of man is a cooking animal," wrote James Boswell in the eighteenth century. Now, humans are animals who neither cook nor necessarily eat. This is the first generation to test what happens to human society when meals are removed from the equation, not because of famine or poverty, but because we feel that food itself has failed us.

CHAPTER EIGHT

THE RETURN TO COOKING

"I'M TEMPTED TO STEP DOWN," A FRIEND REMARKED one Friday afternoon as we sat drinking tea. We were not talking about her job but about cooking. I first met this friend when our oldest children were one-year-olds at the same nursery, smearing paint at the same easel. Now, in the blink of an eye, our boys were both getting ready go to college, and my friend was wearily looking back at all the meals she had prepared over the years.

As a woman who worked, she couldn't quite fathom why it still fell to her to decide the dinner menu for her family of four. More nights than not, it was still she who stood in the kitchen making everything from scratch, alone except for the radio. All that schlepping and shopping, all those last-minute foraging trips for a bulb of garlic, all that headspace taken up with the minutiae of what her children would and wouldn't eat. So many women had already stepped down from the cooking burden, my friend observed, and good for them. Part of her would love to do the same, and yet, for all her annoyance, something still drew her back to the kitchen and the value of a home-cooked meal.

Our eating lives have changed immeasurably in recent years. We snack more, we eat out more, and yet we often enjoy food less. We eat monotonous foods that give us little pleasure (the Cavendish banana springs to mind) and fast foods that are making us sick. We gulp down protein shakes and granola bars. We obsess about food more and ruin our own pleasure at the table by worrying about whether we should only eat things that are clean.

But we can't fully understand these changes without considering another, even more tumultuous revolution, which is the shifting role of home cooking in our lives from an obligatory task to an optional exercise. As Olia Hercules remarked, in food cultures that insisted on three hearty home-cooked meals a day, there was no call for snack bars. Nor would we be so paranoid about whether what we ate was clean enough or pure enough, I suspect, if we were still cooking most of our meals from scratch and knew the ingredients we were consuming. All over the world, old-fashioned cooking skills have been lost in the course of the nutrition transition. The loss of the cooking habit is one of the main reasons we have become such easy prey to the marketing of ultra-processed foods. And yet, as New York cookery writer Deb Perelman has rightly observed, "There are many good reasons to never cook at home."[1]

Unlike any previous generation, we can live our lives—and often deliciously so—without cooking. The home-cooked meal, which used to be a thrice-daily bedrock for many people, has become something we can manage without. With rising prosperity, ever more women take work outside the home, leaving much less time to make the wholesome but laborious dishes of traditional cuisine. A range of products and services have leapt into the vacuum left by women's absence at the stove. Many of these products are highly processed, but more recently, it's become possible to eat home-cooked food even if we are not cooking it ourselves. I spoke to a writer from Turkey who regretted the fact that her daughter, who was in her early twenties and training to be a lawyer, showed no desire to cook. But she admitted that her daughter still ate very well, because she ordered fresh homemade food from one of the new home delivery services that had sprung up in Istanbul.

The death of cooking has been mourned loudly and often. In 2015, in the *Washington Post*, Roberto A. Ferdman announced the "slow but steady disappearance of cooking in the United States." Perhaps the weightiest voice lamenting the decline of cooking has been that of Michael Pollan, who claimed in his 2013 book, *Cooked*, that Americans were cooking less with each passing year. One of the main sources for both Ferdman and Pollan's gloom about the state of modern cooking is a food industry analyst named Harry Balzer who for three decades has been tracking what two thousand households eat as part of a series called "Eating Patterns in America." It's Balzer's contention that cooking is in terminal decline, a human activity that will one day seem as antiquated as "darning socks." "Face it," he told Pollan, "we're basically cheap and lazy."[2]

The data, however, tells a rather different and less depressing story, one in which cooking is not completely dying but slowly being reinvented and maybe even reclaimed—not by everyone, for sure, but by a large enough chunk of the population to represent a significant trend.

In a world where no one has to cook, it is strange and wonderful that so many of us still opt to do it. So far as we can *tell*. Of all the hard-to-measure things about our diets, the extent to which we cook or not is one of the most elusive. It's partly that when we talk about "cooking," we mean such different things by it. Some people think that what they are doing can't be cooking if it isn't an elaborate and hot multicourse meal, while others consider opening a can of beans and slotting a piece of bread in the toaster to be cooking—and who is to say they are wrong? We have all kinds of preconceived notions about what counts as "proper" cooking.

In the end, as with any other aspect of our diets, the best way to measure rates of cooking seems to be to ask lots of people what they do, in as much detail as possible. Our best information on how much cooking people actually do in the United States comes from a 2013 study in the *Nutrition Journal* coauthored by the indefatigable Barry Popkin. The authors used data from six national diet surveys and six time-use studies conducted between 1965 and 2008 and covering tens of thousands of people. The study shows that there was indeed a massive drop in the amount of time the average American spent on cooking from the 1960s to the 2000s—particularly if that person was a woman. In 1965, 92.3 percent of women were regular cooks, dropping to 67.7 percent in 2007–2008. The amount of time the average woman who cooks spent doing so also dropped, from 112.8 minutes in 1965 to 65.6 minutes in 2007. Thinking of my friend and her frustrations, 65.6 minutes a day in the kitchen sounds like plenty.

Like other aspects of our eating, cooking has become a very polarized activity. Half of us cook and half don't. Only slightly more than 50 percent of the US population (including both men and women) spend any time on cooking on any given day, which is certainly a huge decline since the 1960s. But what's really striking is that the percentage of adults who cook isn't actually decreasing, contrary to what Pollan says. As long ago as the 1990s, the decline in cooking started to level off. So far from cooking ever less, the researchers found that the percentage of Americans who regularly prepare food at home has remained pretty stable from the nineties onward.

The story of modern cooking is not a simple tale of decline but something more complex and hopeful. When we say that "no one cooks anymore," we often have in mind a particular version of home cooking that depended on women being confined to a life of unpaid labor. By contrast, the new cooking of our times is done by a wider range of people in a wider range of ways. For hundreds of years, the pinnacle of cooking was supposed to be the invention of a new dish. But perhaps the greatest invention is to find a way of cooking that sits more easily with the demands of our lives.

THE DABBLER COOK

Cooking is in "long-term decline," announced a much-shared article in the *Harvard Business Review* in 2017. The headline declared, alarmingly, "Only 10% of Americans Love Cooking." The article told a story that is now very familiar. For decades, food manufacturers have been trying to persuade us that we are too busy to make our own supper and therefore need their convenient products to save us all the fuss and bother.[3]

Eddie Yoon, the author of the article, is a Korean American consultant who argues that cooking from scratch has been the victim of long-term shifts in consumer behavior. Yoon has spent two decades advising consumer packaged goods firms in the United States. When I catch up with Yoon over the phone from his home in Chicago, he tells me that "the trajectory" is that most people "no longer cook most of the time."

Around 2002–2003, one of Yoon's food industry clients (he can't disclose who) asked him to gather data on consumer attitudes toward cooking. Yoon tells me he set up a survey with a "fairly large sample size," somewhere close to ten thousand. "The sample was broad enough and large enough to be able to say 'this is notable,'" he remarks, excitement in his voice. Based on their answers to a range of questions, Yoon slotted Americans into one of three groups. Group one consisted of people who said that they loved to cook and cooked often. Group two hated to cook and avoided it as much as possible by using convenience foods, delivery services, and dining out. Group three were people who liked to cook sometimes and "do a mix of cooking or outsourcing, depending on the situation." According to Yoon's data, at the start of the millennium, 15 percent of American adults loved to cook, 50 percent hated it, and 35 percent had mixed feelings about cooking and only did it sometimes.[4]

Nearly fifteen years later, in 2017, Yoon conducted the same research for a different client. This time, he found that only 10 percent of US consumers said that they loved to cook, 45 percent said they were ambivalent, and 45 percent professed to hate it. Yoon sees this as a significant shift in attitudes toward cooking. His take is that in a very short space of time, the number of Americans "who really love to cook" had dropped by about a third. It must follow, he argues, that cooking is on the way out, especially for the younger generations. As Yoon sees it, home cooking is on the verge of "category obsolescence."[5]

Yoon's article seemed to confirm what many had long suspected: that for all the foodie obsessions of the modern age, the general population doesn't actually cook very much. The more we make a fetish out of high-end cooking by chefs on TV, the harder it seems to be to do it ourselves. This is what Michael Pollan in 2013 called the "cooking paradox": at just the moment we outsourced most of our meals to the food industry, we suddenly became interested in watching other people cook.[6]

When it appeared in September 2017, Yoon's article spawned a series of the usual despairing responses: "90 percent of Americans don't like to cook—and it's costing them thousands each year," ran one headline.[7] This is not just an American phenomenon, and neither is the despair. I've talked to British schoolteachers who speak of the lost generations of families where neither the children nor the parents know even the basics of meal preparation. The head teacher of one school in the east of England told me about a family whose children kept coming down with tummy bugs. It turned out that the parents didn't realize that raw meat needs to be refrigerated. School cookery teachers speak of pupils who have never handled an onion in its papery skin, let alone chopped it.

Yet if cooking skills are undoubtedly being lost by some households, they are being regained by others. Yoon's research doesn't have to be read in the way that he chose to present it. Let's return to those numbers about attitudes to cooking in the United States. The proportion of people who say that they absolutely *love* to cook may have gone down since the start of the millennium, but the proportion of people who *don't mind* cooking and do it sometimes has, surprisingly, grown substantially. A full 45 percent of people in Yoon's research now describe themselves as occasional or ambivalent cooks, whereas fifteen years earlier it was just 35 percent. When I ask Yoon about this, he agrees. "If you flipped the question and looked at the numbers of dabblers who can cook at least one thing, then yes, I think that's probably going up."

What's striking is not that so many don't cook but that so many still do, even when we don't have to. Another striking change in Yoon's data is the fact that only 45 percent now declare themselves as *hating* cooking, where previously it was 50 percent. This tallies with what we can glean about American cooking habits from national diet surveys over the years. To me, the growth of the dabbler cooks is potentially a huge change in just a decade and a half. This is a phenomenon we don't hear much about: the glimmers of a cooking revival.

Michael Pollan has argued that TV food has transformed cooking "from something you do into something you watch." Yet it's simply not true that the TV cooks of the Food Network and the mountain of culinary videos on YouTube have had no impact on real-life cooking. Many of today's cooks learned what they know from the screen rather than from a mother or grandmother. This may not satisfy our prejudices about how things should be, but it can still result in deliciousness. Restaurant critic Marina O'Loughlin wrote that the most "flawless" Thai sticky rice she ever tasted outside Thailand was served at Siam Smiles in Manchester, a tiny restaurant whose owner, May, learned to cook by obsessively watching YouTube videos after she put her three children to bed. Plenty of nonprofessional cooks also attempt to copy the TV cooks, judging from the rush on ingredients that happens when a much-loved voice recommends a particular item. British cookery writer Delia Smith sparked a frenzied demand for cinnamon sticks and puréed chestnuts after her Christmas cooking shows aired in 2009.[8]

Over the past decade, sales of raw ingredients have finally started to edge back up again after years of being squeezed out of the shopping basket. From 2009 to 2013, the top twenty-five US food and beverage companies lost the equivalent of $18 billion in market share. To the dismay of sellers of packaged foods, some consumers have started avoiding the middle aisles where the ultra-processed foods are and focusing their purchases on the fresh vegetables, fruits, proteins, and cereals at the periphery of the supermarket. The annual volume of packaged food sold in the United States dropped by more than 1 percent year on year from 2013 to 2015. This might not sound huge, but given the sheer scale of the food industry, a single percentage-point drop in sales represents millions of consumers deciding to change the way they feed themselves.[9] Robert Moskow, a Credit Suisse analyst, was quoted in *Fortune* magazine as saying that consumers "have more and more questions about why this bread lasts 25 days without going stale."

Many people who, just fifteen years earlier, had never so much as picked up a pan or a knife are now making cooking a regular part of their lives. These new part-time cooks may not prepare every single meal from scratch. Maybe they didn't learn to make pot roast at their grandmother's side. Sometimes they may go out for a burger or stay at home and order in Vietnamese food. No matter. When they need to, they can get out a chopping board and a knife and make something tasty and edible for themselves.

Here's another kitchen development we don't hear enough about: when it comes to American men in particular, as opposed to women, cooking has been a growth activity over the past fifty years or so. In 1965, according to Popkin's data, only 29 percent of American men made cooking a regular part of their lives, whereas by 2008, 42 percent did. The amount of time a male American cook spends in the kitchen has also increased, from thirty-seven minutes in 1965 to forty-five minutes in 2008. If these men were following Edouard de Pomiane and his ten-minute cuisine, that would be enough time to rustle up four stylish dinners.

After all these centuries of mothers stirring a pot, why shouldn't others have a turn? In the end, my friend realized that what she really wanted to happen in her house was not so much that she should step down from food shopping and cooking as that her husband and teenagers should step up. And they have. The last time we spoke, her husband had started making many more meals, from channa masala to fish pie. What's more, he was enjoying his time in the kitchen as a chance to unwind from work and remove himself from the demands of email.

Eddie Yoon tells me that he too is one of the new generation of male cooks: "I would be one of the 45 percent who likes it sometimes, maybe even loves it." Yoon's Japanese American wife doesn't enjoy cooking at all, and he is the one who prepares most of the meals for their three children. Because of their Japanese Korean heritage, he wants Asian cuisine to be part of his children's education. When we speak, he tells me he has recently been trying to teach his children to enjoy kimchi. "I thought, what if I sauté it in butter?"

I find it strange that a man who conducts these delightful experiments in the kitchen should be so insistent that cooking is on the way out. Like many of us, Yoon seems to be too hung up on a vanished way of eating to recognize the exciting ways in which cooking is being reinvented. When Yoon predicts the death of cooking, what he means is the kind of laborious "scratch cooking" he tells me his Korean mother did every day of his childhood with vegetables fresh from the garden. He does not see himself as a true cook because life gets busy, and sometimes he orders takeaways, something his mother never did.

The real headline here isn't that nine out of ten Americans don't *love* cooking. It's that over half of them—including quite a few men— don't hate it anymore. The death knell for cooking was sounded too soon. Contrary to popular opinion, we are living through a culinary renaissance the likes of which has never been seen. It is happening not in restaurants with three Michelin stars or the mansions of the rich but in modest kitchens where ordinary people are taking ingredients and turning them into vibrant and flavorsome meals. Cooking is no longer a dull duty but something that we actively choose to do, whether we do it to become healthier, to save money, to educate our children about kimchi, or just for the sheer fun of it.

COOKING BY NUMBERS

"When in life do we ever get to see something to fruition?" asks Patrick Drake, the UK head of HelloFresh, the meal-kit subscription service, when I meet him at his offices on a fashionable street in London's Shoreditch. Drake and I are perched on the steps of a small amphitheater decorated with fake grass, a room where HelloFresh UK regularly holds its own in-house version of TED talks. Drake is wearing a beanie hat and nibbling on a packet of mixed unsalted nuts; he looks every inch the former lawyer-turned-entrepreneur that he is. He is trying to explain to me why he believes that, contrary to all the predictions, cooking has the potential to be more popular now than ever.

In working lives dominated by screens, Drake sees cooking as a chance "to create something with your hands and get praise for it." HelloFresh delivers boxes with recipe cards and preportioned ingredients, down to the last clove of garlic or sprig of cilantro needed to make them, so there is no waste—except for the large cardboard box that the food comes in, but at least that can be recycled.

"We do the prep. You be the chef" was a phrase used by Amazon in 2017, as the online shopping business announced it was entering the meal-kit delivery service business in the United States, alongside not only HelloFresh but Blue Apron, Plated, and others. In just five years, the market in meal kits grew from nothing to $5 billion in the States alone. By 2026, the US market is forecast to be worth $36 billion.[10] This is still small fry within a US grocery market worth around $750 billion a year, but considering these kits have been around for such a short time, it's a remarkable shift in cooking behavior.

When the UK version of HelloFresh was launched in 2012— as of 2017 it also operated in seven other countries, including the

Netherlands and France—Drake realized there were many young professionals like him who wanted to make cooking part of their lives but saw recipes as fraught with obstacles. "I tried to make a Vietnamese pho once," he tells me as we stand looking at the four thousand secondhand cookbooks in the HelloFresh library. "It cost me almost forty-five pounds for ingredients that will never get used again." He clearly still feels resentful about the half a kilo of unused cassia bark that remains in his kitchen from that pho.

If you can afford ninety-five dollars for three "Family" meals for four people (or sixty dollars for three vegetarian meals for two), Hello-Fresh will deliver recipes and everything you need to cook them, whether it's a few grams of grated parmesan or six boneless chicken thighs, all neatly packed in a brown paper bag. It's like having your own butler and sous-chef, leaving you with nothing to do but the fun stuff. "We said we wanted to change the way people eat forever," says Drake. The original concept of the meal kit was pioneered in Sweden, where the meals were marketed specifically at families with children, but Drake and his business associates had a hunch that this way of cooking might have a much broader appeal.

Traditional recipes make all kinds of assumptions about our knowledge of food. Chinese recipes may ask the cook to add ingredients to the wok when it is "almost smoking hot." This assumes that someone has already handled a smoking-hot wok and knows what it looks like a few moments before it reaches that stage. Meal-kit recipes, by contrast, assume no knowledge, working on the understanding that most kitchens, even in affluent households, may include little more than a few pots, a knife, and a wooden spoon. "We can never assume the customer has equipment," says Patrick Drake. If a recipe calls for pesto, it will ask for basil and nuts to be chopped by hand, in order not to exclude any customers who don't have a food processor. If something requires rolling, the HelloFresh recipe card will suggest using a wine bottle covered in clingfilm instead of a rolling pin.[11]

As a staunch rolling pin owner, I am not among the target audience for meal kits. I expected to find the whole business patronizing and a bit pointless. After all these years choosing my own produce, trying to source the ripest tomatoes or the freshest fish, I did not think I could enjoy cooking something where someone else had chosen the ingredients for me.

Yet when my family finally tried HelloFresh for a few months, on and off, I was startled by the way that the experiment made us all look at cooking anew. If you asked me, I'd say I love cooking. I've cooked two decades' worth of Christmas dinners for my family. I've steamed artichokes, baked birthday cakes, and known the satisfaction of rustling up soup from a few remnants at the bottom of the salad crisper. Happiness, for me, is a chopping board, a sharp knife, and a lemon. But trying out meal kits revealed to me that my feelings about being the cook of the household were more ambivalent than I had realized. I found myself unexpectedly crying at the kitchen counter when I recognized that, for once, I was not the one responsible for deciding what we would eat for dinner or lugging the ingredients home, because they had arrived packaged on my doorstep like a thoughtful gift.

During the weeks that we tried out HelloFresh, my husband did far more cooking than usual, and it also turned the teenagers in my household into independent cooks. My oldest son, then seventeen, enjoyed meal-kit cooking more than any other kitchen activities I've cajoled him into over the years. As a boy who grew up swapping Pokémon cards, he trusted the reassuring words on the shiny recipe cards more than a nagging parent's voice. His normal repertoire of fried eggs and pasta carbonara suddenly expanded to Thai fried rice with eggplant, shrimp rigatoni, and vegetable stir-fries flavored with hoisin sauce. My fourteen-year-old daughter said that cooking from the recipe cards made her feel like a TV chef.

It isn't that the recipes themselves (at least the ones we have tried) are particularly inspired. In my experience, the range of vegetables—in the HelloFresh boxes, at any rate—was centered too much on out-of-season red peppers and zucchini, the meat was not free range, and the cooking times were often too short. When I tell Patrick Drake that some HelloFresh dishes are not simmered long enough for my liking, he explains that according to market research his customers have a "sweet spot" of about twenty-seven minutes when it comes to the amount of time they want to spend cooking a meal. "If people see on the recipe that it takes forty-five minutes they say, 'I don't have forty-five minutes.'"

In a strange way, the ultra-modern cooking of meal kits recalls the structures of an earlier era, when menus were ritualized through the days of the week. In Italy it is still common to eat like this. At trattorias in Rome, the daily specials are comfortingly predictable, as food writer Rachel Roddy has explained in *The Guardian*: "Friday is the day for pasta and chickpeas or salt cod, Saturday for Roman-style tripe with mint and pecorino, Sunday for fettuccine with chicken livers and then roast lamb, Monday for rice and endive in broth, Tuesday for pasta and beans, Wednesday for whatever you fancy, and Thursday for gnocchi."[12]

Like Soylent and the other meal replacements, the meal kit is a technology for shutting down some of the overwhelming options of modern life. "We want to remove the tyranny of choice," Drake comments. Once your box has arrived, you have no further decisions to make except which dishes to eat on what days. There are no arguments about the menu itself. You know—for example—that Monday will be Mexican tostadas, Tuesday is freekeh pilaf, and Wednesday is herb-crusted fish with green beans and potatoes.

Meal kits are a glimpse of what cooking looks like when it's become entirely optional. It feels like power and freedom, not drudgery and boredom. In the postwar years, cooking and convenience seemed to be mutually incompatible, but meal kits bring convenience and cooking together. To be the person who escapes the demands of phone and computer, if only for twenty-seven minutes, feels like a marvel. It is good for both body and mind to smell the pungent rasp of garlic against a blade, to watch as halloumi cheese takes on a brown rubbery crust, to feel the waxy rubble of chopped pistachios under your knife. I've spoken to friends who say that meal kits have enabled them to take pleasure in cooking for the first time in their lives.

Who would have thought that consumers could be persuaded to pay *more* for a meal that they have cooked themselves? By my calculations, at the time of writing, it costs over twice as much to make a HelloFresh penne all'arrabbiata with smoked pancetta and fresh basil as it would to buy the same ingredients for this fairly basic pasta supper yourself. But you would miss out on the little card telling you that you have consumed three of your five a day of vegetables and urging you to "smile and tuck in!"

AFTER DECADES OF LOSING STATUS TO PREPREPARED food, cooking has itself become aspirational. The rise of meal kits is part of a bigger move back toward cooking. At the university library where I do most of my research, I've noticed that the students have started bringing in their own boxes of vibrant vegetable-based meals for lunch. Sometimes, I hear them swapping vegan cooking tips during their study breaks, something that would have been unheard of a few years ago. Whether because of health or flavor or for the sheer pleasure of it, a significant minority of consumers have returned to the kitchen over the past few years.

Given the choice between cooking and not cooking, many of us now are opting to cook. Today's cooking may not always look much like the cooking of the past, but that is not necessarily to be mourned. By lamenting that only 10 percent of Americans love home cooking, we reveal a shared assumption that cooking is something that we *should* love. In the past, this would have seemed a laughable proposition. Cooking wasn't something you did because you loved it. You did it—"you" almost certainly being a woman—because you had no other option.

In 2016, the food historian Rachel Laudan reflected on the cooking of her mother, who was born a hundred years earlier at the height of the First World War. Laudan's mother was a farmer's wife whose life, like the lives of most wives in the 1930s, was dominated by the production of meals. "Cooking was her job," writes Laudan, "and it was a relentless one. She had to have breakfast on the table at 9, dinner at 12.30 and tea, the last meal of the day, at 5."[13]

It is exhausting just to read about the routine Laudan's mother followed. Each morning she rose and made a cooked breakfast of bacon and eggs, boiled eggs, kippers, or sausages:

> She toasted bread under the grill and put it in the toast racks. Then she flipped the seersucker tablecloth washed to faded rusts and greens on the breakfast room table, placed the pot of tea under its green tea cosy, and added the hot water jug and milk jug on the tray by her place at the head of the table. On went the china, cutlery, a butter dish, home-made marmalade, and the toast racks. Half an hour later she cleared the table and washed up.[14]

All this toil for just one meal. The whole procedure was repeated three more times each and every day. Each time one meal was over, Laudan's mother flipped the tablecloth and started afresh.

Someone once asked Laudan if her mother was a "good cook" and she felt "at a loss" to reply. By today's measures, Laudan's mother's culinary standards were astonishingly high in many ways. "Never do I remember a meal being late, never do I remember a tough pastry crust or a fallen cake." Everything that the family ate was local, fresh, and artisanal. Yet Laudan's mother had no choice about whether to be a good cook or not. It was simply what was expected of her and of every other farmer's wife in England at that time. She did not cook because she loved cooking but because this was the role that life had allotted her.

There was nothing unusual in the way that Laudan's mother cooked. If anything, her life in the kitchen was easy by the standards of the day. At least a farmer's wife had access to plentiful meat and vegetables, whereas city cooks in early twentieth-century Britain were expected to produce the same quantity of meals but with meager ingredients and limited equipment, often in single-room dwellings where there was no kitchen and no escape from cooking. We idealize the homespun meals of the past, imagining rosy-cheeked women laying down picturesque bottles of peaches and plums. But much of the art of cooking in premodern times was a harried mother slinging what she could in a pot and engaging in a daily smoke-filled battle to keep a fire alive and under control, on top of all the other chores she had to manage.

Before we offer too many lamentations for the cookery of the past, we should remember how hard it was—and still is, for millions of people—to cook when you have no choice in the matter.

THE LAND OF COOKS

"The poor don't cook" is the lazy view, much spouted. I've often heard snobbish people say that the poor could afford to eat perfectly well, if only they stopped going to McDonald's and bought frugal staples such as rice and beans. In 2014 in the United Kingdom, the Conservative peer Baroness Jenkin remarked that people were going hungry because they did not know how to cook, pointing out that she had cooked herself up a bowl of porridge for breakfast only that morning that cost mere pennies, so why couldn't the poor do the same?[15]

But data from the United States suggests that people on low incomes actually spend significantly more time cooking at home on average than people on higher incomes. This is not to say that everyone on low incomes cooks every day; as we have seen, when eaters are poor in both time and money, packaged snack foods may look like the answer. Yet from the 1990s to the present day, people on low incomes increased the average amount of time they spent cooking, from 57.6 minutes a day in 1992 to 64 minutes a day in 2007–2008. In the United Kingdom, likewise, women doing low-paid manual work seem to be far more likely to spend 30 minutes or more cooking a day than high-paid women in the managerial and professional class.[16]

A few weeks after my family's trial of HelloFresh, I meet Kathleen Kerridge, a thirty-eight-year-old antipoverty campaigner and journalist. Kerridge has four children, suffered a heart attack at the age of twenty-nine, and was subsequently diagnosed with breast cancer. Her husband was made redundant, and for a while, the whole family of six was managing on a food budget of around fifty dollars to last the whole week. In HelloFresh terms, this budget would buy one and a half family meals for four. Kerridge had to stretch the same money to twenty-one meals for six. She and her husband are both now working in low-paid jobs, but after they have paid for rent and heating and school uniforms, there's still never quite enough left for food, and it takes ingenuity to cook every day, always making do with less than she would like.

Kerridge is driven mad by many of the assumptions made about the way that people with squeezed budgets cook. "They say we are all eating ready meals," she tells me at the conference coffee break, "but I can't afford a ready meal for six people." She has no choice but to cook from scratch every single day, but it is not always a happy experience. "It's awful going 'round the supermarket telling kids they can't have this or that." Unlike many children, Kerridge's have always loved vegetables, but she has to ration how many they can have on her budget because greens just don't fill them up compared to pasta or bread.

For Kerridge, some of the aspirational qualities of modern cooking have only made her life harder. There is an exciting world out there, of shiny-haired people eating shiny vegetables, and she can't afford to buy into it. Her teenage daughters see "courgetti" in the supermarket— zucchini (courgettes) cut into long strands like spaghetti—and beg her to buy it, because they've seen it on Instagram. "But it costs ten times as much per kilo as raw courgettes," says Kerridge.

It's not that it has ever been easy to feed a family on a low income, but in the past, there was a shared assumption that cooking was a difficult duty for everyone. Now, it's different. Kerridge feels taunted by the "ideal world" that cooking has become in our shared imaginations. She finds it heartbreaking, as she has written, to see her children watching her trying to summon up a meal "out of three spring onions and a cup of rice." In an ideal world, writes Kerridge, "I would make my children smoothies every day for breakfast. I would source only the finest ingredients. I would only buy free range organic produce. I would eat 10 portions of fruit and veg a day." In the real world, dinner is whatever she can afford, accompanied, if she is lucky, by "some green beans from the freezer."[17]

So, is cooking power? Maybe it depends on who is doing it. Patrick Drake tells me that HelloFresh has disabled customers who feel deeply empowered by meal kits, because they enable them to cook whole meals for their families for the first time and gain a new sense of independence and respect in the process. But cooking is a different matter when it's the only way you can put food on the table. Both in the rich countries of the world and the poor ones, there are still millions of cooks who produce meals each day for scant reward, except for the cook's honor of nourishing loved ones. These cooks are doing a vital job and producing food that often protects their families from diet-related diseases. Yet somehow the cooks who produce this food rarely get the respect they deserve.

"We fetishize the scent of cardamom in rice pudding but the women producing it are forgotten," Prajna Desai remarks one day from her home in Mumbai. For three and a half months in 2014, Desai ran a cooking workshop in Dharavi, the largest slum in Mumbai, and then documented the cooking and lives of eight of the women she worked with in her cookbook *The Indecisive Chicken: Stories and Recipes from Eight Dharavi Cooks*.[18] It is one of the most original recipe books I have ever read. She found that, without recipes to aid them, these women cooked with "an amazing intersection of rote thinking and confidence." These were cooks who were willing to take risks and did not mind if a dish took longer than twenty-seven minutes from start to finish. They hand-rolled rotis from pearl millet and bathed white fish in a sauce of tamarind and fresh cilantro. There was subtlety and quick-wittedness in their food, such as green bitter gourds stuffed with a sweet peanutty filling, tied with string and shallow-fried.

Desai went into Dharavi in 2014 and asked for volunteers for a cooking project. At the first meeting, there were around thirty-five women—"I think they wanted an excuse to leave the house"—but only eight of them kept coming back day after day. She met the group of women, many of whom didn't write, three times a week, and they recorded one another's recipes. To start with, it was a struggle for her to persuade the women that they had anything worth sharing. "They thought they had come to learn recipes rather than being the teacher," Desai recalls.

In the spring of 2017, I catch up with Desai over Skype. She is at home in Mumbai, in a room whose walls are lined with colorful textiles, with the sound of birdsong in the background. Desai, whose background is as an art historian, was first approached by an arts NGO to do a project in the Dharavi slum about art and health. It occurred to her—as it has occurred to many others before—that cooking is itself a form of art, but not one that is given its full due.

As Desai sees it, the economic success of India depends on the artisanal brilliance of women like the ones she met in Dharavi. "The whole country would just fall apart without the unpaid labor of cooks who produce meals three times a day. Cooking every day is still very common in India," she tells me. Yet the wife and mother who produces exquisite meals on a shoestring budget is still not accorded high status for her work.

One of the cooks Desai worked with was Kavita Kawalkar, a twenty-five-year-old housewife training to be a teacher. One of Kavita's dishes was a fish masala consisting of fish in a spicy tomato sauce with cilantro. Before she embarked on the sauce, Kavita took a whole onion and a hunk of dried coconut and charred them separately on two gas burners. As Desai writes, "This important first step sweetens the onion and persuades the hard coconut to release its sweet oil." The finished dish, Desai found, tasted like "pure velvet for the tongue."

When we lament that hardly anyone in the West cooks anymore, the kind of cooking we have in mind is exactly what millions of wives in India still do every day. The eight Dharavi women in Desai's project make everything from scratch, even elaborate flatbreads. The food they produce is varied, delicious, and mostly very healthy (although Desai did notice that some of them used huge amounts of cheap cooking oil, part of the global trend that we saw in Chapter 1). A typical meal might be rice, lentils, some kind of fresh chutney or salad or other homemade condiment, one or two curries, and always homemade yogurt as an accompaniment. This feat of skill and organization is repeated three times a day, yet the women Desai worked with did not see themselves as "someone who has something to contribute."

In pure food terms, what these cooks are doing is beneficial. But Desai worries about the human cost—across India and in other traditional Asian societies—of this laborious meal production. The women she met in Dharavi produce super-varied meals not because they want to but because "they are expected by their families to cook a certain variety of dishes." Cooking is nonnegotiable, and being obliged to do it makes these women feel less important than the men they cook for.

The extraordinary meals produced by traditional cooks in India and elsewhere deserve to be celebrated. In our age of diet-related ill health, these cooks are serving up preventative medicine three times a day. It would be a huge waste if such culinary traditions were to die out at just the moment we need them more than ever. The question is whether some version of this cooking can be preserved without forcing women to lead lives that few would opt for, given the choice. Can we retain the benefits of traditional cooking without forcing anyone to endure the lowly status of a traditional cook?

NEW KITCHEN RULES

The morning I meet Yemisi Aribisala it is a rainy day in Cape Town, and her Uber is stuck in traffic. I wait for her in a hipster café, nursing a streaming cold and sipping a flat white. On arrival, Aribisala, who has a deep, infectious laugh, hands me a bag of pungent delights, with scents so forceful they pierce through my nose. Here are tiny dried fish and a tub of smoky chili paste, honeybush tea, black dried figs, and an assortment of impossibly strong-smelling spices unlike anything I have ever seen or smelled before.

Aribisala, a Nigerian food writer, has presented me with what appear to be half the contents of her kitchen. She has done so with pride. To Aribisala, Nigerian food is an unknown treasure, if you can call something eaten by 180 million people unknown. "I tell people that the world has not met Nigerian food," she announces at the start of her 2016 book *Longthroat Memoirs*, an account not just of Nigerian cooking but of what it means to be a woman who cooks in the twenty-first century.

A highly educated woman, Aribisala has written that some of her female relatives say she should not waste her brain spending so much time in the kitchen. But she has a deep attachment to the Yoruba gastronomy she grew up with. Aribisala's version of cooking is not the easy HelloFresh way, but a series of complicated steps that would have once been considered "culinary tests of womanhood," such as handling a pot on a cooktop without oven gloves and skillfully cutting up yams with no chopping board. She prides herself on being brave enough to dip her hand into a pan of hot water to retrieve a hard-boiled egg. Like Eddie Yoon in Chicago, she keeps the traditions alive for her own children: mucilaginous soups thickened with okra and tough, full-flavored old chickens cooked with Scotch bonnets and nutmeg.

It's not hard to see why many people hate cooking in this day and age, Aribisala concedes. She knows many affluent women in Abuja who feed their children instant noodles "when their N60,000 a month cook-steward takes his day off. They order pizzas and Chinese takeaway for lunch and dinner." Opting to be a noncook is a perfectly understandable position when the Nigerian housewife who cooks—unlike the male TV chef—is given such feeble recognition. The female cook is still expected to feed everyone else in the family before herself, to endure burns and cuts and "calloused hands," and hardly to be thanked.[19]

But despite all this, Aribisala maintains that the kitchen of a Nigerian woman is a "room of power and illumination," a place where people arrive hungry and leave fed. It is for this reason and many others that Aribisala refuses to stop cooking, even though she technically could choose not to. She is not the only one refusing to get out of the kitchen. "Many Nigerian women," she explains, "cook their own meals at weekends, freeze them, thaw them according to meticulous timetables after long work hours during the week. At night after transits in snaking traffic, they arrive home and wear the apron and the kitchen with pride."[20]

Given that we can put food on the table through such a variety of other means, it is striking that so many of us—whether in Africa or North America or Europe—are still choosing to cook. Modern cooking may not be much like the cooking of our grandparents, but that is partly because modern households are not the same either. In some states in the United States, twenty-one out of every thousand households are now headed by a same-sex couple. These households are remaking the rulebook on who cooks, who cleans, and who looks after any children. The upshot may be a renewed respect for the value of cooking, because there is no assumption that just one person has a duty to do it.[21]

We had it wrong all these years when we thought that to be a cook somehow wasn't powerful. What could be more powerful than engaging in an activity that supports the health and well-being of your loved ones (including yourself)? This can be seen when we look at the consequences of the nutrition transition around the world. A very clear line can be seen between local communities' abandonment of home cooking—rice and beans in Mexico, vegetable soup in Portugal—and the rise of diet-related diseases.

We should offer more appreciation, as Aribisala writes, for "how hard it is to put a pot on a fire and produce something edible and creative every day without any glory in sight." We also need to find ways to make cooking much—*much*—easier. Most of the "easy cooking" promoted on the internet is anything but.

One answer to the conundrum of how to keep cooking alive in the modern age is to become less rigid about what cooking actually is. It's still home cooking if a man does it. It's still cooking if you take yesterday's homemade curry and reheat it in the microwave and there is only one of you at the table to enjoy it. It's still cooking if you use half a dozen "lazy" kitchen tools and shortcuts your great-grandmother never dreamed of.

Across the world, cooks are turning to fresh ways to make the same delicious meals. In Brazil, the gadget of choice is the pressure cooker, which turns out the old feijoadas and black bean stews in a fraction of the time and using far less fuel. In the United States, United Kingdom, and India, it is the Instant Pot, which has developed cult status among busy people (I include myself). This electric gadget, which combines a slow cooker and a pressure cooker with a rice cooker and several other functions, can enable you to make delicious slow-cooked soul food without the slow hours of work. In Italy, it is the Bimby (known elsewhere as the Thermomix), which can not only blend like a food processor but can also cut, mix, weigh, and stir and make a risotto so suave and comforting that no one eating it would know that a strong-armed mamma hadn't stirred it lovingly for twenty minutes. One in thirty Italians, including many men, now owns a Bimby, a startlingly high number considering that it costs around a thousand euros (more than a thousand dollars).[22]

Cooking has always been a trade-off between the demands of our lives and our desire for delicious and wholesome food. That trade-off has never been more complicated than it is today, but where we find ourselves now isn't all bad. To cook when we have no choice about it is one thing. But to choose cooking out of all the other demands on our time and attention is a much more positive act. It is a gesture of every-day determination and of love, whether you are cooking for yourself or others. You don't have to take a whole head of cauliflower, split it into an orchard of tiny white florets, and roast them in a hot oven before dousing with lemon, but you do so anyway. No one forced you to take corn and potatoes and leeks and cream and cook them into a rich and soothing chowder, but still, you are glad that you did.

But the old ways of eating haven't entirely vanished yet, and we are reminded of this every time we cook, even if it's only a bowl of soup for one. The act of taking ingredients and applying heat to them has a kind of deliberateness that so much of our other eating has lost. It forces us to pay attention to ingredients, if only for a moment on the way from hand to pot.

So much has changed so fast in how we eat; yet food is as central to our lives now as it ever was. To cook is a way to honor that fact and escape some of the contradictions and excesses of our food culture. We may not cook every meal or every day, but to cook at all is an antidote to so much of the craziness of modern life. The emails may go unanswered. Spin class may be forgotten. The WhatsApp notifications buzz away. Leave them. Dinner is served.

CHAPTER NINE

CROSSING THE BRIDGE

MUCH OF THE WORLD HAS NOW REACHED THAT FAIRY-tale ending, "And they never went hungry again." As we've seen, today's plentiful food supply is a dream by the standards of earlier generations. For centuries, our hungry ancestors consoled themselves with stories of gingerbread cottages and magic porridge pots that would never dry up and roast Christmas goose—the fattier, the better. The old European fairy tales offer a dream of sheer quantity: of carbohydrates in your belly and the impossibly sweet sense of security that comes with not going to bed hungry.

The stories we tell about food matter because they affect what we eat and how we eat it. "And they never went hungry again" was the right story to get us from stage three to stage four of the nutrition transition: from receding famine to abundance. From the year 2000 to 2015, the level of hunger in developing countries fell by 27 percent. That is nothing short of miraculous. It remains a cause of vast collective shame that more than 800 million in the world are still going hungry, including millions of children under the age of five who will be left permanently stunted by lack of food early in life. In the face of such horror, it was perfectly natural to focus on the urgent question of simply generating enough to eat. In 1941, Franklin D. Roosevelt declared that the "freedom from want" was one of the four basic human freedoms.[1]

Our food system was bequeathed to us by mid-twentieth-century plant breeders who were obsessed with the idea of food yield, above and beyond quality or diversity or sustainability. Norman Borlaug, father of the Green Revolution, is said to have saved a billion lives with his invention of semi-dwarf, high-yield wheat varieties. Those numbers are hard to argue with. Susan Dworkin, a writer who started off working at the US Department of Agriculture, has described the mindset of the seed breeders she used to work with. They would ask, she recalls, "How much food could you get out of an acre? How many people could you feed? That's where they are. That's what they think. They are not looking at the dinner table. They're looking at the swollen belly."[2]

But gazing around today's baffling and contradictory food supply, "they never went hungry again" no longer looks like the right happy ending to be telling ourselves. There are new diet-related horrors, such as the rising number of two-year-olds who are having all their teeth extracted because they consume so many sugary drinks or the adults having limbs amputated as a result of type 2 diabetes. Our agricultural system is still optimizing for productivity and yield even though the problems facing most people on the planet are now a combination of overconsumption and undernourishment.

"And they never went hungry again" does not sound like an unhappy ending for the thin-fat babies of India or for those time-pressed and low-income consumers in the West who only avoid the empty stab of hunger by eating a nutritionally unbalanced diet of packaged snack foods. This is not a story that redresses the monocultures of the Global Standard Diet or the fact that most diets now are far too rich in sugars and refined oils and not rich enough in micronutrients. This simplistic story does not save us from the misery of eating disorders. There are food hungers, moreover, that cannot be met by nutrients alone. A hunger for more time to eat and cook in our busy working lives. A hunger not to be obese. A weary desire not to feel so guilty and agonized about our food choices all the time.

An attachment to the mindset of never feeling hungry again explains much (though not all) of our irrational behavior around food. When eating, we are often scared of the wrong things. We look with terror on small portions and treat anyone who suggests we might eat less of one of our beloved treats—from meat to sugar—as the wicked witch in a fairy tale. Conversely, when we see an overfrosted, intensely sweet five-hundred-calorie cupcake covered in sprinkles and twice the size of the cupcakes of the 1980s, we may see only joy and celebration and happy endings.

We need new ways of thinking about food to help us adjust to the abundance we now find ourselves surrounded by and to start to build a better way of eating. Given how often humans have changed their culture of eating in the past, there is every reason to hope that the strange place we now find ourselves in with respect to food is not the final chapter but a phase that our great-grandchildren will one day look back on with distanced bemusement, just as we look back on driving cars before seatbelts or the way that cigarettes were once advertised as a cure for asthma. "What! Parents fed their children rainbow-colored sugar-frosted loops for breakfast? And the packet was legally allowed to claim that it was a 'smart choice'?"

I remain hopeful that we can somehow fight our way through stage four of the nutrition transition to stage five. Stage five, explains Barry Popkin, would be a new pattern of eating and living that would come about through "behavioral change." It would be a change in culture as much as anything. As in South Korea, stage five would still give us the affluence of stage four but would express it with a more restrained way of eating centered on vegetables. Personally, I would keep the kimchi and the exciting new cooking of today's food culture but ditch the clean eating and guilt. Stage five would see a return to water as the drink of choice and a reduction in the consumption of caloric beverages. There would be an increase in "purposeful activity" as people lessen their dependence on the cars of stage four and start to walk and cycle again. The hope is that after the huge acceleration in diet-related disease of stage four, stage five would see a reduction in NCDs and in obesity. My own hope is that we would also see less fatphobia and a realization that there is more to healthy eating than Instagram bodies. Food would once again become something that nourishes us rather than makes us sick.[3]

We will not reach this state, however, without outside help, which means that governments and other organizations will have to do their bit to reset the needle on food. Changing the world in which we eat will require action on multiple fronts, from agriculture and better regulation of food markets to education and cooking lessons. As food journalist Felicity Lawrence wrote in 2018, it becomes ever clearer that reshaping this environment requires government intervention. "Although we can exercise a thousand rebellions individually," wrote Lawrence, "none of us can change the landscape on our own."[4]

THE SWEET GREEN GRASS

For some reason, even in the midst of this epic crisis of diet-related ill health, the idea that government has a duty to help its citizens eat and drink more healthily remains deeply controversial. Any time someone proposes a law to reform the food environment, there will be angry cries of "nanny state!" In 2012, there was an outcry in New York City when Mayor Bloomberg proposed a limit on the size of sugary soft drinks sold, with the portion of the largest bottle capped at sixteen ounces. To be clear: Bloomberg's proposal was never intended to be a ban on soda, nor was it an exercise in rationing. A citizen would still have been free to buy as many sixteen-ounce bottles of soda as he or she could afford. Yet Bloomberg's law—which was repealed in 2014 by the New York City Board of Health—was widely attacked as a paternalistic violation of personal freedom. "He seems to believe that the best, highest purpose of government is to tell people how to live their lives down to the most minute detail," observed one critic of the mayor.[5]

The widespread resistance to increased government intervention on food is partly, I suspect, down to a very human impulse; from childhood onward, we hate being told what to put in our mouths, and it puts us in a rebellious mood. To be told what to eat can feel like an intrusion, as any parent feeding a picky child soon discovers. "Oh, take a hike!" I often think when I see those government-approved leaflets on how to eat that advise that a rice cake with cottage cheese would make a fine snack. And I like cottage cheese.

There's a second reason why the idea of food regulation is so despised in our times, and that is the psychological problem of "less." In the past, we were more content for governments to intervene to protect the food supply on our behalf because this protection largely took the form of making sure that we had enough to eat. From ancient times, it was a basic role of the state to try to ensure that there were full granaries and enough nourishment to go around. Even wartime rationing was as much about making sure that each person had his or her share as it was about requiring people to cut back. My mother, born in 1941, still goes teary-eyed with gratitude when she remembers the free black currant syrup that the British government gave out to children under the age of five during the war, to ensure that they had enough vitamin C.

The dilemma today is that governments and cities can't simply keep promising us more food, because on average, we already have far too much. As a result, most forms of government action on food feel punitive. Bloomberg saw his law as simply doing his duty to protect the public health of a city where an estimated 987,000 people suffer from type 2 diabetes. But to its critics, the law felt like stealing food from our plates (or rather soda from our bucket-sized cups).

BUT WHEN IT COMES TO FOOD POLICY, THERE IS ONE thing that governments not only could but should attempt to offer people more of, and that is food quality.

For decades, we have been bamboozled by quantity and have paid far too little attention to the quality of what we eat. Hungry? Eat more. Obese? Eat less. Whether we are trying to lose weight or gain it, there has been a narrow and false view of nutrition as simply a matter of calories in and calories out. But a healthy diet is not just a question of quantity, and it can't be defined only by what it *doesn't* contain. As we've seen, a health-giving way of eating is not simply a diet that is low in sugar and fast food but one that is high in such nutrient-dense foods as yogurt, fish, nuts, legumes, and green vegetables.

There is still a remarkably common view that good-quality food is something that can only ever belong to the posh or the rich. Radio broadcaster Derek Cooper once wrote that what was wrong with food culture in the United Kingdom was the existence of two different kinds of food. The first was the cheap and nasty stuff, which Cooper called "tins of highly coloured rubbish." The second was expensive foods that attract words such as *real*, *natural*, *organic*, *traditional*, *pure*, and *handmade*. "But," Cooper asked, "shouldn't all food be as safe and as pure and as fresh as possible? Why have cheap bad food at all?" Over the past couple of years, those in food development circles are finally starting to ask this question of quality too. The global food system now needs to address not only hunger but also the provision of better-quality diets across the board, for rich and poor, young and old.[6]

I was trying to imagine how to tell this new story about food quality when suddenly I remembered "The Three Billy Goats Gruff," which used to be my oldest son's favorite story when he was little. It's about three brother billy goats: little, medium, and big. They live in a field of brown scrubby grass and yearned to eat something better. Across the river is a meadow with the sweetest, greenest grass that any of them had ever seen. But before they can get there, the goats have to cross a bridge where an ugly troll lived who would gobble up anyone who dared to cross. Luckily, the goats manage to trick him into not eating them, and they make their way safely across the bridge to the sweet green grass.

"And they reached the sweet green grass" is a better food aspiration, it strikes me, than "and they never went hungry again." This is a story about food quality, diversity, and enjoyment as well as quantity. It is not just any old grass that is at stake; it's the good stuff. (I always picture a lush meadow with many types of wild grass.) The story can be read as being about reconnecting with the land and recognizing the goodness in real food when we see it, as our hunter-gatherer ancestors did. Unlike clean eating or so many of the other extreme food reactions of the modern world, this is a story about running toward food, not away from it. It isn't about guilt or denial but better nourishment. We are the billy goats gruff, and whatever our size, we need to cross the bridge uneaten. But there are ugly trolls lying in wait for us as we cross. The trolls take many forms depending on your situation in life. The question is how we get past these predators and reach the sweet green grass: a way of eating that actually gives us both pleasure and health.

I can't pretend to know what lies over the bridge. Some say the future of food will entail eating insects, and others, that it will be algae. By definition, the change to our diets that is so urgently needed can't be just one thing because as we have seen, the way a society eats at any given time relates to multiple other things too, including politics and economics, education, and patterns of work. My hunch is that the sweet green grass will look very different in different places, as we start to reclaim and reinvent some of the local foods that our globalized food system has done so much to obscure. We need to go back to asking what the land nearest to us can best provide in the way of food rather than expecting farmers to produce an identikit series of items the world over.

The change that we need is a transformation in culture as much as ingredients. Whatever stage five ends up looking like, the first step is to change the stories we tell ourselves about food. And the next is to put a foot on the bridge. Around the world, there are a few hopeful signs that this is already starting to happen.

KILLING THE CARTOON CHARACTERS

When I ask Barry Popkin whether anywhere in the world gives him cause for optimism that there could be a shift away from our current destructive patterns of eating, he initially responds with one word: Chile.

As of 2016, Chile had the highest average consumption of sugar-sweetened beverages on the planet. The country was also eating unusually large quantities of salty snacks and chips and packaged sweet desserts. More than half of the food purchased by the average household was ultra-processed, and Chileans had the second-highest rates of obesity in Latin America, after Mexico. According to estimates by the Chilean Ministry of Health, around 66 percent of Chilean adults were overweight or obese, when as recently as the 1980s it was more common to be malnourished. So far, so familiar. All the Latin American countries have suffered the worst effects of the nutrition transition later than in the United States or Europe but at an accelerated pace.[7]

The difference was that as of 2016, Chile also enacted the most aggressive range of laws against unhealthy foods that the world has yet seen. The government passed an 18 percent tax on sugar-sweetened sodas, one of the highest sugar taxes to date. But this was only the start. In an audacious move that has been welcomed by public health experts but described by an industry spokesperson as "invasive," the new Chilean food laws of 2016 forced cereal manufacturers to remove all cartoon characters from boxes. This was an attempt to change the basic symbolism of packaged food. There would be no more cute bunnies and polar bears on boxes of sugar-frosted cereals and, there-fore, no more illusions that such items were an essential element in a happy childhood. In 2018, a headline in the *New York Times* summed up the law as "Chile Slays Tony the Tiger."[8]

Removing cartoon characters from cereal boxes is just one element in a portfolio of food laws that, taken together, constitute an all-out assault on an obesogenic food culture. The guiding force behind these regulations is Senator Guido Girardi, a former children's doctor who has described the sugar in junk foods as the "poison of our time." Girardi has been campaigning for stricter food laws in Chile since 2007, but the passing of the laws kept being delayed by corporate interests. Girardi describes the eventual triumph of Chile's strict food laws as "a hard-fought guerrilla war." Schools in Chile are no longer allowed to sell ultra-processed foods such as chocolate or potato chips. In or out of schools, the sale of Kinder Surprise eggs has been banned, because the toy inside is seen as a kind of enticement to eat sugar.[9]

The most striking aspect of the Chilean food laws has been the new food labeling requirements. Nutrition labeling has been used many times before in efforts to persuade populations to adopt healthy diets, but most of these efforts have been, frankly, useless. By listing grams of fat, say, or portion sizes on a packet, the idea is that consum-ers will be armed with all the tools they need to make healthier food choices. But this is to assume that consumers are rational, educated individuals who choose their food freely, without constraints of time or money. It is also to assume that they care as much about health as a nutritionist does.

Thus far, most food labels have been a spectacular failure in the fight against diet-related illness, because the messages are too confus-ing or too subtle and take no account of the way that people actually behave in a supermarket. Often you have to squint at the writing to figure out what the portion size really is and what, precisely, the color-coding system is trying to tell you. Numerous studies have shown that nutrient lists on the front of a pack are more likely to be used by high-income consumers who already have healthy-eating preferences. Most food labels are thus entrenching the very health inequalities that they are designed to redress.[10]

Chile decided to make its food labels much clearer and starker. It started in 2014 with a series of warning labels on children's foods such as flavored milks and highly sweetened yogurts and breakfast cereals. Simple hexagonal labels announced "warning: high in sugar," "warning: high in salt," "warning: high in saturated fat," and "warning: high in calories." The Chilean government was alerting its citizens about the ugly troll. By the standards of US food labeling, the messages were astonishingly blunt. Marion Nestle, professor of food policy at New York University, hailed the new rules as "absolutely astounding." But for Senator Girardi and other Chilean public health lobbyists, the labels still did not go far enough. Girardi called them "rubbish" because they did not cover a wide enough range of foods, in his view, and because manufacturers were allowed to print the warnings in red, blue, or green—colors that have positive associations for many consumers.[11]

Since then, the labels have been redesigned in black, and they now apply to thousands of products in the average Chilean supermarket. These dark warnings are plastered not only on cookies and snack foods but on many items that have long marketed themselves as healthy to consumers, such as yogurts, bottles of light salad dressings, undiluted juices, and granola bars. It is too soon to tell yet whether the labels will help to reverse obesity rates in Chile, but there are signs that consumer behavior is already changing, as people are slowly nudged away from certain foods. Nearly 40 percent of Chileans surveyed said that they now use the black logos to help them decide what to buy. Patricia Sanchez, a mother of two, told a reporter from the *New York Times* in 2018 that she never read food labels in the past, but these new ones "kind of force you to pay attention."[12]

At the time of writing, a handful of other countries have copied Chile in passing legislation requiring warning labels on food. Peru,

Israel, and Uruguay have all followed the Chilean model, and there is talk that Brazil and Canada may follow suit.

The new food laws in Chile have not been universally welcomed. In the summer of 2017, I met a lovely woman from Santiago, a passionate cook, who lamented the fact that her favorite sea salt was now branded with a black label, as if a pinch of sea salt in a pot of soup were equivalent to a portion of french fries doused in salt. Others say that it is wrong to demonize sugar and salt—or any food, come to that—because we risk making people feel anxious about their food choices.

There is no denying, however, that the new laws have spurred the food industry into action. As many as 20 percent of all food products for sale in Chile—more than 1,500 items—have been reformulated in response to the law, with sugars and fats reduced, in order that foods can avoid the dreaded black labels. Coca-Cola has said that 65 percent of the drinks it sells in Chile are now low sugar or reduced sugar beverages.[13]

In a way, this kind of regulation makes the food industry's job simpler too. Food policy expert Corinna Hawkes told me that she meets many representatives from multinational food companies who say they wish they could remove some of the sugar from their products but find it hard to do so, because if they unsweeten their cereals or yogurts too much, consumers may switch to one of their competitors' products. Across-the-board labeling and taxation of the kind seen in Chile creates a level playing field in which no single manufacturer is punished for removing sugar.

Reformulating products to lower-sugar versions has been the default industry response wherever sugar taxes have been introduced. I must confess that lower-sugar ultra-processed food still doesn't look quite like the sweet green grass to me. Several long-term studies involving large cohorts have suggested that low-calorie sweeteners are still associated with greater levels of type 2 diabetes and weight gain. But even if these reformulated foods are not perfect, it is welcome that the food industry is being forced to think harder about the healthfulness (or otherwise) of its products.[14]

Latin America is starting to look like a world leader in food reform. In Brazil, Carlos Monteiro has said of modern diets and the harm they cause that "doing nothing is no longer an option." Monteiro—a professor of nutrition and public health—helped devise Brazil's official nutrition guidelines, which have been praised as the best in the world. Rather than lecturing the population on which nutrients to consume and in what quantity, the Brazilian guidelines talk about eating in terms of real food and the meals that humans eat. They advise choosing mostly "minimally processed foods," but they also recognize that oils, fats, sugar, and salt can be part of "diverse and delicious diets" when used in cooking. The official guidelines celebrate the sharing of food at a table as a "natural part of social life."

There has also been much excitement in nutrition circles about Mexico—where more than 70 percent of the population is overweight or obese—and its sugary drinks tax, which was the model and inspiration for most other sugar taxes, including the one introduced in the United Kingdom in the spring of 2018. The Mexican sugary drinks tax imposes just one peso per liter of sugary drink, equivalent to a few cents. Industry critics of the tax—introduced on January 1, 2014—say it has so far done nothing to reduce levels of obesity or diabetes in the country. This is true, but given that the tax was only introduced so recently, it's a bit soon to judge.

What is clear is that the Mexican tax has triggered a radical shift in drinking habits in a country that drinks more soda than almost anywhere in the world. Researchers at the University of North Carolina analyzed the effects of the tax on purchases. They found, encouragingly, that—based on store purchase records for nearly seven thousand households—there was a 5.5 percent drop in purchases of sugary drinks the very first year the tax was introduced. The second year, in 2015, sales dropped still further—by 9.7 percent. The highest reductions came from households at the "lowest socioeconomic levels"—the very people who suffer the highest levels of diet-related illness. Meanwhile, purchases of untaxed drinks, principally bottled water, went up.[15]

Some argued that the Mexican sugar tax would not be high enough to have a telling impact. But the data suggests that even small adjustments to the price of food and drink can create significant changes to consumer behavior, particularly when accompanied by new messages about food. As in Chile, the soda tax in Mexico has been part of a wider campaign to reduce the positive associations that junk food and soda have in the public mind. In 2012, Bloomberg Philanthropies gave money to Alejandro Castillo, a consumer advocate who had long been fighting to keep junk food out of Mexican schools. Castillo staged a series of events in which he presented the cartoon characters on sugary foods as the "Junk Cartel"; Tony the Tiger was characterized as "el Tigre" or "the Lord of Sugar." Castillo also produced a series of ads aired on Mexican TV called "12 spoonfuls," based on the amount of sugar in a bottle of soda. One of the posters showed two children with a bottle of soda being thrust at them. "Would you give them 12 spoonfuls of sugar?" the poster asked. "Then why would you give them a soda?"[16]

The true radicalism of these Mexican and Chilean food laws, and the campaigns surrounding them, has been to start to change collective ideas of what is normal when it comes to eating. Imagine being a Chilean child born today who will never associate chocolate cereal with adorable cartoon animals. Deprived of its cartoon mascot, a packet of highly frosted cereal is revealed for the nutrient-poor and boring breakfast that it really is: nothing but a soggy pile of sugar-corn drenched in milk.

"IF I HAD MATHS RESULTS LIKE THIS I'D BE OUT OF A JOB"

Transforming an entire food environment is a daunting task—and it also seems a far-fetched scenario in most countries, whose governments are reluctant to do anything to antagonize the food industry. It is easy to look at the sheer scale of the problem—as with climate change—and become defeatist. But maybe we don't have to change the whole food environment to achieve beneficial changes in the way people eat. Maybe, for starters, we could change just one small corner of it.

What we eat is hugely shaped by the environment we happen to live in. But the environment doesn't just mean the world at large. Every day, we make our food choices in a series of smaller micro-environments. A micro-environment could be anything—a shop, a restaurant, a food court, a family dinner table. Theresa Marteau is director of the Behaviour and Health Research Unit at the University of Cambridge, where she is leading a four-year research program to test ways of redesigning some of these micro-environments in such a way that people are nudged automatically to eat and drink in a healthier way. She is interested in such questions as how people can be persuaded to eat less meat and what are the most effective ways to get people to reduce their portions.[17]

Marteau and her colleagues have found that seemingly tiny adjustments to the setting in which we select food and drink can have a dramatic impact on the choices we make. For example, by changing the size of wine glasses, people can be nudged to drink more or less. The obvious implication is that to persuade populations to moderate their wine consumption, one answer would be for bars and restaurants to use smaller glasses (if only the bars and restaurants could be persuaded to sign up for it). This sounds so simple, I say to Marteau. "Sure! Absolutely!" she exclaims. The beauty of this approach, she points out, is that no one would be telling anyone else how much to drink. Marteau's decades of work on food and behavior have taught her that people profoundly dislike being told how much to consume. But when our micro-environments are adjusted, we change our behavior quite painlessly.[18]

"Choice architecture" is the term that behavior scientists now use to describe the design of environments in which all the choices are healthier ones. I find this an appealing idea: building a house of good choices where our food desires are free to roam wherever they please. It's very different from the way we live now, where our good intentions crumble in the face of micro-environments crammed with cues to overeat.

If we are likely to change our food habits anywhere, it is as children, at school, surrounded by our friends. This is why it is so sad that so many schools in the United States have felt pressured into allowing branded junk foods into the cafeteria. In 2018, the schools of Houston signed a new four-year $8 million contract with Domino's, which has created a special version of pizza for schools called "Smart Slice," in which the ingredients are tweaked to comply with school food standards. In 2016, according to Bettina Elias Siegel, an advocate for better school food, Smart Slice was being sold in six thousand US school districts in forty-seven states. Siegel points out that, even if Smart Slice is healthier than a regular slice of Domino's pizza, its presence in schools creates brand recognition and thence brand loyalty in children. Basically, these companies are using the school cafeteria to teach children that junk food is normal.[19]

YET SCHOOLS CAN ALSO BECOME ENVIRONMENTS IN which children learn healthier food habits than at home. In the spring of 2017, I visited Washingborough Academy in Lincolnshire, a state-run primary school where the entire school day is focused on encouraging the children to build a healthy relationship with food. When the head teacher, Jason O'Rourke, arrived at the school, one of his first actions was to get local businesses to sponsor him to plant an apple orchard with as many local heritage varieties as he could find. Every class in the school has its own kitchen garden, where children grow vegetables that they use in cooking classes and to supply the school cafeteria. As I toured the school with O'Rourke, we paused in a classroom where children were earnestly debating whether a red pepper was a fruit or a vegetable. We stopped off at the school "snack shack" where another group of children were preparing sweet potato muffins to sell at break time for a healthy snack. We saw children pulling cabbages from the ground and others chopping organic leeks to make a leek and potato soup.

By the standards of most schools in Britain, Washingborough devotes a huge amount of attention to food. O'Rourke, who might be the most unusually food-minded head teacher I've ever met, is motivated by his horror at the high rates of child obesity in the United Kingdom. Almost one in five children in Britain arrive at school at age four already overweight or obese. By the time they leave primary school at the age of eleven, it's one in three. As O'Rourke dryly remarks, "If I had maths results like that, I'd be out of a job." What really shocks him is that the British education system does not currently regard child health as particularly important. He has decided that at Washingborough, at least, the children will grow up learning to eat in a new food culture, one that prioritizes both pleasure and health.

I got to know O'Rourke in 2016 because we are both involved in a group trying to bring to the United Kingdom a new system of food education called TastEd. It's based on the Sapere system of sensory food education that has been used in Sweden and Finland (as well as France and other countries) for more than twenty years. The idea is that a child learns best about food not through lectures on healthy eating but through his or her senses. In a typical TastEd lesson, children might put on noise-canceling headphones and try "loud" and "quiet" foods (noisy celery, silent strawberries). Or they might smell different spices hidden in jam jars and try to identify what they are.[20]

O'Rourke is excited about the possibilities of TastEd because he feels that so many children who arrive at Washingborough are lacking in even a basic knowledge of flavor and texture. A few months before I visited the school, the Washingborough chef had added a dish of roast chicken legs to the menu. The chef was upset that more than half the portions were returned to the kitchen uneaten. It turned out that even though chicken was the favorite meat of the children, most of them had never before encountered a bone. The only chicken they knew was boneless and bread-crumbed, in nugget form. The teachers had to coax the children into tasting this dark, slightly chewy meat, reassuring them that yes, it really was chicken.

This is just one school in just one corner of England, but O'Rourke sees his job as promoting a more positive food culture than that espoused by the outside world. Within the school boundaries, he and his teachers can act as architects of the children's food choices. O'Rourke often remarks that when it comes to food, it can be much easier for teachers to have a positive impact than it is for parents, for whom the feeding relationship is more emotional. At most schools, when a child has a birthday, the whole class celebrates with sugary cake and cookies. At Washingborough, as a birthday treat, the child donates a book to the school library with a special bookplate inside it containing the child's name.

Changing the food culture in even a single school will have a ripple effect that reaches far beyond the classroom. At one of the first TastEd sessions on a Friday in the summer term of 2017, Washingborough teachers asked six- and seven-year-old children to look at different kinds of apples and describe their colors and shapes: red or green, shiny or dull, round or squashy. Then they tasted them. On Monday morning, several parents reported that when they went grocery shopping that weekend, their children begged them to buy different types of apples as fervently as they might once have clamored for sweets.

CELEBRATE WITH OLIVES

Most of our habits of eating are established in childhood and once learned become extremely difficult—although not impossible—to unlearn. Looking at the figures on children's changing diets and weight around the world, it's not easy to summon up much optimism that our future relationship with food will be healthier than our past one. The vast majority of countries are experiencing a shift from child obesity to more child obesity, from junk food to more junk food. But there is at least one place in the world where the number of children living with obesity is finally coming down. It has happened thanks to a coordinated effort from teachers and parents, politicians and healthcare workers, social workers and psychologists, sports coaches and dietitians, and even supermarkets and fast-food chains.

Childhood obesity is declining in Amsterdam, a city of almost a million people. From 2012 to 2015, the percentage of children in this Dutch city who are overweight or obese went down 12 percent, from 21 percent to 18.5 percent. This decline included some of the city's poorest children, those from mostly immigrant families who have traditionally been more overweight than children from white Dutch families. The change did not happen by accident. It was the result of a remarkable initiative, the Amsterdam Healthy Weight Programme (AHWP), whose ultimate goals are that no child in the city will have an "unhealthy upbringing" and that by the year 2033 all of Amsterdam's children will "have a healthy weight." Based on what the program has achieved so far, that goal does not look as far-fetched as it did when the program started.[21]

We hear so much about responsible consumption, and usually it is a code for saying it's someone else's problem. "Drink responsibly" is the slogan on advertisements for hard liquor, as if being an alcoholic were a form of naughtiness rather than a disease. Being overweight is often seen—by those fortunate enough never to have struggled to lose weight—as a failure of personal responsibility. The single biggest flaw with this argument is children. How can a child be considered responsible for her own obesity when she has virtually no control over the foods that she is fed?[22]

Consider the case of Ruth, a fourteen-year-old morbidly obese Surinamese girl living in Amsterdam. Ruth is in danger of liver failure because of her weight. In Surinamese culture, a larger body size in women is perceived as beautiful, but given that Ruth is growing up in the Netherlands, surrounded by messages about thin-body ideals, she is depressed and hates the fact that she cannot fit into her jeans. This is a problem afflicting many of the Surinamese and Turkish children in the city, who also tend to live in the poorest neighborhoods. Ruth's parents are separated, and she is not given the opportunity to eat healthy foods at either parent's house. Her father is a taxi driver who is hardly ever at home to oversee Ruth's meals but also does not allow her to go out alone. Without some kind of outside help, the chances of Ruth "taking responsibility" for her health and finding a different way of eating would be negligible. But now, thanks to the AHWP, she is one of the children in the city who have been targeted to receive intensive counseling as well as other forms of help, personalized to an individual's needs, ranging from access to sports clubs to courses on healthy eating and shopping.[23]

There are now children with obesity like Ruth in every city in the world, and except for their parents, it can feel as if no one cares about how hard life is for them. We see people like Ruth and we think: not our problem. The example of Amsterdam offers hope that this laissez-faire mindset may yet change. The program recognizes that these children need help with their mental as well as their physical health.

When the Amsterdam program was set up by the city council in 2012, the aim was to change ideas of what is normal for child health. Compared to the rest of the Netherlands, the city had an unusually high percentage of overweight children (21 percent as compared to 13 percent), and the city council had had enough. Eric van der Burg, the deputy mayor for healthcare, pushed through a series of no-compromise reforms, including bans on fast-food advertising at sporting events. The only drinks that children can bring to Amsterdam schools are now water or unsweetened milk—no juice—and treats such as cake and chocolate are banned from school premises. To start with, teachers found that many parents had to be convinced that juice and fruit drinks were not healthy, because many of them had previously believed these sugary drinks to be full of fruit and vitality. Five years into the program, however, the idea of satisfying thirst with water is becoming normal again, helped by the fact that the city has installed fifty additional water fountains.[24]

The program works on many fronts at once, because it starts from the premise that childhood obesity "is a wicked, complex problem": not the fault of any one person or factor. In social science, so-called wicked problems are those that appear impossible to solve, because they are of such complexity that there seems to be no fixed stopping point. With a wicked problem, there is also no obvious single point at which to intervene, because the problem intersects with so many other problems. But the AHWP shows that even the wicked problem of obesity can be ameliorated.[25]

One of the main ways that the program is delivered is through schools, where the mission is to make exercise and healthy foods a normal part of every child's routine. The city leaders aim for all Amsterdam schools to be healthy, but around 120 of them have registered to be special "jump-in" schools that place an extra emphasis on health, from physical activity to eating. In the old days, children often brought sugary cakes and sweets in for birthdays, and the Amsterdam parents—like parents everywhere—sometimes competed to offer the most generous treats because they felt like an extension of love. Now, there is a rule in jump-in primary schools that birthdays are celebrated with healthy foods such as fruits and vegetables. A popular birthday option now is to bring in a selection of vegetable skewers to share with friends, consisting of tomatoes, cubes of cheese, and green olives. Celebrate with olives![26]

The messages Amsterdam children hear in school are reinforced in other areas of their lives. The marketing of unhealthy foods to children around the city has been banned, and meanwhile, dozens of food businesses have joined the Healthy Amsterdam Business Network. If a child should walk into a branch of McDonald's near one of the jump-in schools, he or she is not allowed to buy anything except an apple without the permission of an accompanying adult. Community leaders talk to parents about the importance of children getting enough sleep and about eating family dinner together. The program also supports families who lack the resources or education to make so-called healthy choices. Sometimes, it's not even possible to start addressing the problem of a healthy diet until these families have been given assistance with other, even more pressing concerns, ranging from lack of money to decent housing.

Amsterdam shows what can happen when we stop seeing poor diet as a question of individual willpower and start addressing the underlying causes. A report from the program coordinators in 2017 found many things to be proud of, especially the fact that it had succeeded in reaching children in the poorest neighborhoods. Many of the targets set in 2012 had been met. There had been a "significant reduction" in the number of obese five-year-olds, and four out of five neighborhoods where the program had been launched had seen the number of overweight children fall. Healthy living is now considered an aspect of urban design in Amsterdam, where the idea of "built-in exercise" has become a key element in any future city planning. The organizers of the program detected a paradigm shift in the way that child health was discussed in the city. They had noticed a swing from people saying that "parents and children are responsible" to an attitude of "we are all responsible."27

For centuries, the size of a city was limited by the number of citizens it could feed. Cities in turn have intimately affected the way that we eat. In modern times, both cities and suburbs became environments that could have been designed to make people fat. Every day, we are navigating landscapes swamped with foods engineered to be overconsumed and streets so cluttered with cars that it is hard for anyone—least of all, a child—to ride a bike or walk. Amsterdam points the way to a different kind of city, one whose shared sense of prosperity will be judged on how well it looks after its children.

THE JOY OF GREENS

Everyone knows that one of the biggest problems with modern eating habits is that in most countries, most people don't eat anything like enough vegetables. This is often talked of as an intractable problem, because vegetables are just so, well, unappetizing. But what if that could change? What if we could learn an appetite for greens?

Unlike the South Koreans or the Portuguese, the British have not always had the happiest relationship with vegetables. Earlier generations tended to boil them to a gray and flavorless oblivion, sometimes with a pinch of bicarbonate of soda for good measure. During the decades after the Second World War, the British ate a surprisingly large amount of vegetables, but more from duty than from pleasure. Veg was a nonnegotiable element in the "meat and two veg" that formed the basic structure of food on a British plate. You ate the vegetables, but you didn't actually expect to enjoy them. That was what meat and potatoes and gravy were for. Survey data from 1958 suggests that an average British adult ate around fourteen ounces of fresh vegetables every day—which is close to the current official recommended daily intake set by the World Health Organization.[28]

During the postwar years, the tradition of "meat and two veg" in Britain started to die out, at least as a daily occurrence. It was displaced both by a host of new convenience foods and by a less rule-bound eating culture. As soon as the British lost the sense of social obligation to eat vegetables, greens started to vanish from our plates. Parents who had suffered the torment of being forced to swallow over-boiled leeks in white sauce were not going to inflict the same pain on their own children. As of 2017, the average British adult eats a mere four and a half ounces of vegetables a day, little more than one and a half portions.[29] The second-most-eaten source of vegetables for a British child is canned baked beans in sugary tomato sauce.[30]

Over the past decade, in Britain as elsewhere, there has been a revolution in the image of vegetables, helped by chefs and cookbook authors such as Yotam Ottolenghi, whose recipes made cauliflower seem like just about the most exciting thing anyone could eat. Through learning to love other cuisines, greens have become an object of desire, from the pak choi of China to the cavolo nero of Italy. For many of us in modern Britain, vegetable eating has become a magical kind of game: gotta catch 'em all! We visit farmer's markets and try beautiful varieties we never had to suffer in the school cafeteria: candy-striped beetroot, Romanesco cauliflower, orange tomatoes, yellow zucchini.

The worry, however, is that the people in Britain who need it most are not sharing in this vegetable revolution. Some in the United Kingdom now eat a whole rainbow of vegetables, while others eat fewer than ever. "People have very high awareness about eating vegetables," says Anna Taylor, executive director of the Food Foundation, an independent body that tackles healthy food policy, "but that doesn't translate into consumption." It was concern at Britain's low vegetable intake that led Taylor to launch Peas Please, an ambitious, independently funded three-year project that is working with eighty different organizations—ranging from growers to hospitals and supermarkets to caterers—to make it easier for everyone to eat veg, on all incomes. Taylor sees eating more veg as the single diet intervention that could have the most impact on both health and the environment. "It's no good if it's just purple carrots in a farmer's market," says Taylor.

In today's liberated society, vegetables will return to our plates only if people actually want to eat them. The aim of Peas Please—which Taylor admits is a "massive, massive challenge"—is to get everyone in Britain eating an extra portion of vegetables every day. A big element in the Peas Please campaign has been to raise money for an advertising campaign that will attempt to make vegetables cool. The Food Foundation analyzed how much advertising money was spent on different foods in the United Kingdom from 2010 to 2016 and found that the proportion of all food and drink advertising spent on vegetables was just 1.2 percent. By contrast, the amount spent on advertising cakes, biscuits, confectionery, and ice cream was high and rising: from 18.8 percent in 2010 to 22.2 percent in 2015. In 2015, the total amount spent on advertising vegetables in the United Kingdom was £12 million, compared to nearly £87 million for soft drinks alone. It is extremely hard for vegetables to compete for a place on our plates when so much money is being spent to make junk foods seem more lovable. By crowdfunding for a new series of vegetable adverts from concerned private citizens, Peas Please hopes to create a more level playing field between broccoli and potato chips.[31]

Another reason that so few vegetables are eaten in Britain—and elsewhere—is that they are hardly promoted in the supermarkets where most people buy their food. Peas Please has extracted pledges from mass market businesses such as Sainsbury's supermarket, which has promised it will always include vegetables in the store's "fresh inspirational plinths" (a fancy way of saying the shelves at the end of the chilled aisles). Sainsbury's has also experimented with offering an in-store "vegetable butcher." Customers pick whichever loose vegetables they fancy, then take them to a counter where someone will cut them any way you please: julienned, wavy-cut, ribboned, or spiralized, at no extra cost. Another business to pledge is Greggs, a low-cost chain of cafés that promises to sell an extra fifteen million portions of veg between 2018 and 2020 in its sandwiches and salads. Taylor tells me she deliberately approached Greggs because she wants to drive change in vegetable eating "beyond the metropolitan elite."

As in Amsterdam, the Peas Please project is working on many fronts at once. For too long, the question of whether or not someone eats vegetables has been seen purely as a matter of individual choice. But when you live in a society that treats cabbage as a bad joke and where fresh vegetables are not seen as a basic component in a meal, we are beyond the territory of personal preference. Solving the problem of the British and greens requires a "food systems" approach, remarks Taylor. Among other interventions, the project is working to get more fresh vegetables into the neighborhood convenience stores where many low-income families do their grocery shopping.

One of the inspirations for Peas Please was the Healthy Bodegas scheme in New York City. In many neighborhoods of the city, the main source of food is a bodega or convenience store. Traditionally, these stores centered on ultra-processed foods with a long shelf life, such as canned soup and potato chips. But in 2005, the New York City Department of Health launched a scheme to get healthier items into the bodegas. As of 2012, the scheme had worked, with more than a thousand stores in which the owners were supported in selling a range of wholesome fresh foods at low cost, especially fruits and vegetables.[32]

Low-income families are just as likely as anyone to enjoy vegetables if the obstacles to eating them can be removed. In 2013 and 2014, the charity Alexandra Rose gave eighty-one London families vouchers for free fruit and veg at local markets through children's centers in Hackney. Vegetable-cooking classes were offered to help those who wanted them. The benefits of the scheme were not just a short-term nutrition boost but a longer-term change in habits and tastes. The parents commented that they were now "experimenting with different foods—like beetroot." Thanks to the Rose vouchers—which are now distributed in Hammersmith, Lambeth, and Fulham as well as Hackney—families have found themselves choosing to buy veg they had never tried before. One mother said that both she and her child were now less fussy about vegetables. "Now I crave salad rather than kebab," said another.[33]

Peas Please is an attempt to drive this new appetite for vegetables across the whole country. There are already small glimmers of a new vegetable love emerging in the United Kingdom and beyond. In

2016–2017, sales of beetroot in the country grew by £34 million, an increase of 6 percent year on year. Supermarkets have also been taken aback by the success of vegetables sold as low-carb pasta alternatives, ranging from "courgetti" to cauliflower couscous to slices of butternut squash "lasagna." The astronomical sales of these products suggest that vegetables have the potential to reach new markets beyond the elites.

To date, the British conversion to greens is far from complete. It remains to be seen whether the United Kingdom—or the United States, for that matter—will ever have a fully vegetable-centric food culture such as that of Vietnam or India. The problem is one of chicken and egg. The vast majority of fresh produce in supermarkets is grown for yield and uniformity rather than flavor, and so it doesn't taste very good (as well as being less rich in nutrients than the vegetables of the past).

Most Western consumers, traumatized by bad memories of overboiled veg at school or at home, don't yet expect vegetables to taste very good, so we don't complain about the blandness of carrots or the dull sponginess of a supermarket zucchini. But how will we ever get the tastier veg that might convert us to its charms unless we learn to ask for something better?

RECIPE FOR A SEED

The trouble with most recipes is that they start at too late a stage. If you consider a typical American recipe for roasted pumpkin or winter squash, it involves a series of maneuvers designed to correct for a fundamental lack of flavor in the produce itself. Before it is roasted, the recipe may recommend heavily oiling and salting the squash and lacing it with maple syrup, all to disguise the fact that the underlying quality of the orange flesh is so watery and bland.

But what if the recipe started not with a list of ingredients but with a seed? This was the revolutionary idea of American chef Dan

Barber, whose new company Row 7, founded in 2018, aspires to change the way that vegetables, and ultimately other crops, are bred. In 2009, Barber cooked for a group of plant breeders at his restaurant Blue Hill at Stone Barns in New York State. After dinner, Barber took one of the plant breeders, Michael Mazourek, on a tour of his kitchen, where one of his cooks was preparing butternut squash. Barber asked Mazourek, "If you're such a good breeder, why don't you make a butternut squash taste good?" Barber wanted to know why plant breeders didn't shrink a squash down, to make it less watery and more concentrated in flavor. Mazourek, an associate professor in plant breeding and genetics at Cornell University, replied that in all his years working in breeding, "no one has ever asked me to select for flavor."

That one conversation between Barber and Mazourek may just change the future of farming. At the time that Mazourek met Barber, he had already started work on a miniature squash but was finding it a "hard sell because they don't fit a lot of people's concept of what a good squash should be." After his meeting with Barber, Mazourek went away and spent the next couple of years creating a squash—the Honeynut—which in a remarkably short space of time has altered the entire market for squash in the United States. The Honeynut has proved that consumers will pay more for a smaller but better-tasting vegetable, given the opportunity. The Honeynut squash was not seen in a farmer's market until 2015. By 2017, it was being grown in 90 percent of squash farms in the Northeast and was being sold by the million in Trader Joe's and Whole Foods and through Blue Apron meal kits. Dan Barber's ultimate aim is that the Honeynut—and equally tasty iterations of other vegetables—should be available to consumers everywhere through mass-market supermarkets such as Walmart.[34]

The Honeynut is smaller and denser than the average butternut. Depending on what else you are eating, a single one may serve for a portion. They are thin-skinned, do not need to be peeled before eating (so there is less waste), and they pack in three times the amount of beta-carotene compared to a regular squash. The biggest difference of all is that they are so rich in flavor, as if they were self-seasoned, needing nothing to be added. When Mazourek had nearly finalized his chosen seed, he asked Dan Barber back in to cook some of the possible candidates. Barber found that the Honeynut was so flavorsome, it didn't need brown sugar or maple syrup to enhance it.[35]

Barber himself sees the triumph of the Honeynut as an inversion of the values that have driven most of modern agriculture. He has said that when Mazourek first took "this shrunken butternut squash" to "the big agribusiness people" describing the incredible flavor, he was met with a series of rejections. One of them told Mazourek that it didn't fit with the "logarithm" of a squash in the supermarket because it was the wrong size. Another said that it was impossible that consumers would pay 10 percent or 20 percent more for a squash that was 60 percent smaller. "Both," Barber has remarked, "were exactly wrong."[36]

In its focus on flavor and ripeness, the Honeynut squash stands in marked contrast to the way that almost every other commercial vegetable is now grown and marketed. Butternut squash, for example, is always picked green and unripe, because this means that there is no risk of it rotting on the vine. It helps the distributors and the retailers to pick it green, but it doesn't help consumers because we end up eating squash that is always a little bit unripe. By contrast, Mazourek bred the Honeynut so that the skin would turn in color from green to pale honey at just the moment that it is ripe, and the squashes are picked ripe. As Barber has said, "90 percent of the brilliance of this squash is that it's freaking ripe. We don't eat ripe squash. That's crazy."

It is Barber's hope that the Honeynut is just the first step in what he sees as a wider project to "democratize flavor," in making delicious-tasting vegetables available to as wide a number of people as possible. Unlike so many other modern seed varieties, the Honeynut is not patented but open to anyone who wants to grow it. Barber's company Row 7—the first seed company ever built on collaboration between chefs and breeders—has now developed seven varieties of vegetable seeds (and is working on more). These seeds include a pepper, a potato, a "bold and complex" cucumber, and a beetroot that even a child could like.[37]

Developed by a plant breeder in Wisconsin, this is called Badger Flame Beet. The breeder and Barber started having a conversation about why so many people don't like beetroot. The breeder noted that beets contain a compound, geosmin, that provides the earthy, cooked-blood taste that some enjoy but many people (including my children) despise. The breeder decided to, as Barber puts it, "select against geosmin" until he ended up with a beet so sweet, pure-tasting, and lacking in earthiness that it can be eaten raw. Barber sees Badger Flame Beet as "a gateway drug" into enjoying regular beets.[38]

Part of what makes Barber's project so brilliant is that instead of telling eaters that they are wrong not to like squash or beetroot, he is meeting people in the middle, giving them a reason to overcome their prejudices. This approach is more carrot and less stick. The project does not look backward to traditional ingredients but forward to something even more delicious. If this is the future of food, maybe it need not be so bleak after all.

Many attempts to change our diets for the better have paid far too little attention to the role of appetite. Over the past two years, as part of TastEd, I've done a series of sensory food workshops in a local primary school with four- and five-year-olds. Children are often made to feel that they are wrong not to enjoy certain foods, but what if the fault is not with the child but with the food itself and the way in which it is offered? It strikes me that children often have very good reasons for not enjoying fruits and vegetables, because so many mass-market vegetables don't taste very good. Many of the children tell me that they hate tomatoes because they are cold and watery, but when I present small, sweet, ripe tomatoes at room temperature with a concentrated flavor, many self-proclaimed tomato-haters have been converted.

If we are going to cross the bridge to a better way of eating, we need a much broader and more affordable conception of healthy eating, one that welcomes people and their appetites in rather than shutting them out. As we've seen, the nutrition transition has involved a global change of preferences, with the arena of pleasure largely colonized by the sellers of ultra-processed foods and fast foods. But it doesn't have to be this way.

In Washington, DC, dietitian Charmaine Jones finds that many of her clients, who are African American, find it difficult to move away from eating so much fast food because they perceive much of what is labeled as healthy food to be "white people food." Most of her clients are low-income black women suffering with type 2 diabetes, high cholesterol, or heart disease. Much as they want to be healthy, it is not easy for them to change the way they eat when the surrounding culture presents them with a vision of healthy eating that feels exclusive and alien to them: a Goop vision of wellness. One of Jones's clients was Tanisha Gordon, a thirty-seven-year-old IT worker. She was hooked on a diet of burgers, tacos, and fried chicken, which tasted like home to her, even though she was prediabetic. Gordon told a reporter for HuffPost that much of what was sold by mainstream culture as healthy food felt like "a white person's food" to her: "These extravagant salads with all these different ingredients in it, like little walnuts and pickled onions." She was only able to change her diet once she could find a version of healthy eating that spoke to her at the level of appetite. Jones teaches her clients how to make healthier versions of soul food, dishes that provide comfort as well as nutrients.[39]

This kind of positive cultural change will only start to translate across whole populations once governments recognize that access to high-quality food—by which I mean food that sustains health as well as food that people want to eat—is not a luxury but a need. In 1948, the Universal Declaration of Human Rights named food as a basic human right: "Everyone has the right to a standard of living adequate for the health and well-being of himself . . . including food." The problem is that for the decades since, fixated with the specter of hunger, we have allowed our standards of adequate food to be set far too low. Just because someone is not outright starving does not mean that they are adequately nourished. As epidemiologist Dariush Mozaffarian has observed (on Twitter), "Fed but stunted and diabetic is not a great win."[40]

Over the past couple of years, food policy experts have finally started to focus on strategies to encourage higher-quality diets for all, whether it's those suffering malnutrition in poor countries or people with obesity in rich and middle-income countries. As we've seen from Fumiaki Imamura's work, what we *do* eat matters as much for diet quality as what we avoid eating. This is why some African countries have patterns of eating that are the healthiest in the world, because these are among the few places where varied whole grains, vegetables, and legumes are still normal components in everyday meals rather than weird special ingredients that you buy for a weight-loss diet.

Bringing the focus back onto the quality of our diets would require us to think about how to encourage people to eat more of certain foods, as well as less of others. This should mean not just telling us what to eat but helping populations to change their preferences toward healthier diets. It is estimated by scientists at Tufts University that more deaths are caused every year in the United States by diets low in nuts and seeds (59,374 deaths) than by diets high in sugary sodas (51,694 deaths). Yet how often do you see a public health campaign promoting the joy and benefits of hazelnuts?[41]

As the city of Amsterdam has shown, the problem of modern diets won't be fixed by just one intervention. A smart and effective food policy to address obesity would need to work on multiple fronts at once. It would seek to create an environment in which a love of healthy food was easier to learn, and it would also reduce the barriers to people actually buying and eating that healthy food. It would find a way to remunerate farmers for producing better food rather than just more food and a way to rearrange our cities so that healthier food becomes the easy option. We need economic policies that make healthy food more affordable. It also wouldn't hurt if our workplaces recognized that it's not a sign of weakness to stop for lunch.[42]

You did not choose the environment in which you learned to eat, nor did you design the shops in which you buy your food. If you eat too much sugar and refined oil, that says less about you than it does about the world you are eating in. There will always be those who insist that eating is purely a question of personal responsibility and that it is wrong to meddle unduly in food markets. But this is to assume that the food environment we are living in now is natural and normal—as if the presence of sugary drinks and sweets in every store were simply an act of God. In fact, as we have seen, almost everything about the way we eat now is very new and would have seemed strange a few decades ago. There is therefore every reason to think that our food habits can and will undergo another transition.

We never snacked like this and we never binged like this. We never ate such flavorless bread or so many bland bananas. We never consumed such a narrow range of ingredients, on average, across the world. We never ate out so much and we never ordered in so much. We never had so many superfoods or so many french fries. We never cooked so many portions of chicken. We never went on so many diets, and we never drank so many gallons of cold-pressed organic green juice. We were never quite so confused about food and what it actually is.

But if history teaches anything it is that we won't always eat in the particular ways we do now. Here is the consolation of eating in these strange times: the best of it is better than anything that came before and the worst of it won't stay the same forever. Wherever you are and however you eat right now, I hope you can reach your own sweet green grass.

EPILOGUE: NEW FOOD
ON OLD PLATES

I'VE SAID THAT HOW WE EAT IS NOT A QUESTION OF PERSONAL choice. But that doesn't mean it's not worth trying to change the way you eat for the better. As we work and wait for a new and improved food culture to emerge (which may be a long wait), there are still things that we as individuals can do to enjoy the best of modern food without getting swallowed up by it. This kind of personal change can sound trivial, but it isn't. Sometimes it is the only change available to us. I hope you don't mind if I offer some thoughts on how to navigate this world of choice. Nothing is as personal as the question of what we put in our mouths, so feel free to ignore anything that doesn't apply to you.

EAT NEW FOOD ON OLD PLATES

We can't go back to the meals of our great-grandparents, but we could use their plates (and bowls and glasses). So much of our eating now is done without any plates at all and therefore without any sense of occasion. We eat out of cardboard boxes using fingers or plastic forks, which are then discarded (the forks, not the fingers). But the new ways are not always the best. Ceramic plates are a wonderful technology. They are not just reusable and durable but lovable. A good plate—whether it's decorated in blue and white or plain white—gives structure to a meal and can lift the spirits when you eat.

Whenever you get the chance, eat and drink from ceramic and glass rather than from plastic. Not only is it better for the environment; it is better for you. Think back to the Japanese American men in San Francisco who suffered heart problems when they adopted fast American ways of eating. The rituals of eating matter, and a meal eaten from china sitting down is more restorative than one grabbed out of a wrapper. Life does not allow us to eat every meal from ceramic, but the more we can reinstate it as the norm, the better.

Why old plates? Mainly because they were smaller. Fast food is not the only kind of food to have become supersized in recent years. Our lovingly home-cooked dinners have also inflated in size, along with the size of our dinner plates. A large dinner plate in the 1950s had a diameter of twenty-five centimeters, whereas today the diameter of a normal dinner plate may be as large as twenty-eight centimeters—a great expanse of china—with the result that we unthinkingly serve ourselves larger portions than we need.[1] What looks like a generous and heaping portion on a 1950s plate looks measly on a modern one.

The inflation in the size of wine glasses has been even more dramatic. Theresa Marteau, director of the Behaviour and Health Research Unit at the University of Cambridge, found that there had been a sevenfold increase in the size of average wine glasses in England from 1700 to the present day. A typical wine glass in 1700 was a tiny goblet containing just 70 milliliters whereas the average wine glass for sale in 2016–2017 was 449 milliliters. Even allowing for the fact that most people don't fill a 450-milliliter glass to the brim, this is a startling increase.[2]

Once you start using old plates and old glasses, I have found, you begin to reset your instincts about how much to eat and drink, even when you are out of the home, eating from different plates. Many of us have completely lost any sense of what a normal portion looks like—which is hardly surprising given that packaged foods send us such strange mixed messages about how much to eat. Often, the suggested portion on a packet of breakfast cereal will be minuscule, to make the contents sound more slimming than they are. In 2010, a survey of nearly 1,500 elderly South Koreans found that they still had a remarkable degree of shared knowledge about portion sizes, because they all still ate the same traditional cuisine. They all "knew" that a portion of white rice was two and a half ounces and a portion of spinach was one and a half ounces.

Instead of buying into the latest expensive diet craze, head to your nearest garage sale or secondhand shop and buy the oldest and smallest plates you can find. There are plenty around, going for a song. These plates are your friends, because they will give you the portion control that you need, without counting calories or cutting out food groups. Another strategy would be to buy small dishes from a Chinese supermarket and adopt the Asian ritual of eating family style, with several large dishes on the table from which each person helps themselves to a small portion until he or she is full. Or you could buy an Indian thali set, consisting of a ten-inch stainless steel plate with a series of smaller bowls to go on top, a form of tableware that forces you to eat meals that are not just portion controlled but varied.

Bowls are another way to go. Since 2016, bowls have strikingly gained in popularity at the expense of plates, especially among modern cooks who use them to make bowl food: wholesome but delicious assemblages of grains and stews or brightly colored vegetables.

Onto these small plates or into these bowls, you should pile all the joyous new foods that this world has to offer. Use the plates as a way to think about the architecture of a meal and which foods you want to give priority to. If a meal isn't a slab of meat surrounded by a few vegetables, then what is it? Maybe it is a larger amount of vegetables, seasoned with smaller amounts of meat or fish. Or maybe it is three different vegetable and legume dishes of contrasting flavors with some

noodles or grains or flatbreads and pickles. The truth is it can be anything, so long as you find it flavorsome and filling. Try new fruits and
eat spices your grandparents never knew, but don't be scared of carbs
or fats. The plates will see you right.

DON'T DRINK ANYTHING "LIKE WATER" UNLESS IT IS WATER

Some health gurus say, simply, "Don't drink calories." In a world
containing cappuccino and red wine, this seems a little unrealistic.
"Don't drink calories" also leaves open the possibility of downing a
whole two-liter bottle of artificially sweetened diet soda, which carries its own risks. But until human biology becomes more accurate at
registering the calories from liquids, a good blueprint is to approach
drinks other than water—including juice—as if they were snacks, to
be taken sparingly. Ask yourself: Am I thirsty or hungry, or do I have
some other reason for wanting a drink? If you are thirsty, nothing
improves on water.

The best way to avoid drinking fizzy drinks (and Frappuccinos
and related drinks), I have found, is to unsweeten your palate until
the concept genuinely no longer appeals. The same goes for sugar
in hot drinks. Like changing any other habit, it takes time for your
palate to adjust, and you feel wobbly in the early stages. But if you
can make it through to the other side, you find you have changed so
much that you can't imagine swallowing something so sweet. In an
ideal world, our governments would do more to protect us and our
children from the demonstrably harmful effects of sugary drinks. In
the meantime, a dislike for soda and a thirst for water are among the
most useful preferences anyone can acquire.

One exception is tea. The explosion in Chinese obesity coincided
with the sudden displacement of unsweetened green tea by other
drinks. Tea in all its forms—from green to black and caffeinated to
herbal—might be the perfect solution to the question of how to enjoy

the varied drinks of modern life without consuming excess quantities of sugar or sweeteners. Even black tea with a splash of milk has very few calories compared to milky coffee, and depending on how long it is brewed, the caffeine hit is gentler—or nonexistent, if you drink herbal teas. If you think herb tea is boring, take either a bunch of fresh mint or a piece of fresh-grated ginger, steep it in hot water for three minutes, and then strain into a china mug for a drink that is both uplifting and soothing.

Another exception is homemade flavored waters. I've noticed that these are already becoming popular in gyms and hotels, where the water coolers are often strewn with cucumber or citrus slices, to give the water a hint of flavor without sugar or sweetener. I met a Turkish chef who said he made what he called "everlasting water." He chopped up a certain kind of Turkish persimmons and covered them in water to infuse, straining it off over several hours. He told me that the persimmon flavor was faint at first but became stronger over time, imbuing the water with the sweet ripeness of fruit.

DEVOTE LESS ATTENTION TO SNACKS AND MORE TO MEALS

Snacks would never have become such a big part of our eating lives if we hadn't started to neglect our basic meals. Instead of obsessing about the perfect one-hundred-calorie snack, we should spend more of our time planning delicious hearty meals, with gaps in between. Incidentally, on the subject of snacks, Professor Corinna Hawkes has remarked that most packaged snacks such as popcorn that are repackaged as healthy have "absolutely no value" in the diet, consisting as they do of nothing but refined oils and refined grains. As we saw in Chapter 1, one element of the nutrition transition involves the downgrading of bread as a staple food, yet bread is better and more filling than nine-tenths of what is marketed as a snack. "Have a piece of whole grain bread if you want a healthy snack," says Hawkes.[3]

CHANGE YOUR APPETITES

It is easier to change the foods you enjoy than it is to force yourself to eat what you don't like. We often talk as if pleasure is a problem—as if we could only ever enjoy packaged foods high in sugar, oils, and salt. But food without enjoyment is hardly food at all. The single most effective tool against living in an obesogenic world is to work on developing new preferences until you can take pleasure in foods that are actually good for you.

So many of our problems with eating come down to the fact that the foods we feel we *should* eat and the foods we want to eat are two different things. Cauliflower feels like coercion; chocolate feels like love. This is no wonder, given that we are blitzed with the aggressive marketing of foods high in sugar, refined fats, and salt, which sends us the message that these foods are deeply pleasurable. But what if cauliflower could become something to crave? The example of South Korea shows that it's much easier to eat healthy food when—like kimchi—it is an object of desire.

SHIFT THE BALANCE

Don't worry too much about whether you are eating a perfect diet (not least because the pursuit of perfection can make the experience of eating so unhappy). Instead of drawing absolute lines, focus on shifting the balance of what you eat in a healthier direction. For most of us, a less meat or less sugar diet is easier to achieve than one without any meat or sugar at all. The same goes for ultra-processed food. It's hard to avoid altogether, but most of us would benefit from eating less of it. As Fumiaki Imamura's study shows, it is the pattern of eating that matters, but the hard part is that the world around us now sends very unbalanced messages about what is normal when it comes to food. There is a big difference between a diet where more than half of what you eat is ultra-processed (as is the case for the average person in the United States and United Kingdom) and a diet that is just 20 percent ultra-processed.

What does a healthy pattern of eating look like? Many nutritionists advocate the Mediterranean diet, consisting of olive oil, fish, nuts, vegetables, legumes, and fruits. Others prefer the newer concept of a Nordic diet, a sustainable way of eating rich in berries and dark grains such as rye, barley, and oats; rapeseed oil; and oily fish such as herring and salmon. But those of us who live neither in the Mediterranean nor Scandinavia may have to invent our own patterns of eating. Fumiaki Imamura told me that since moving to the United States and Britain from Japan, he had asked many people what a healthy local diet looked like "and no one has been able to answer me." The fact that no one can yet identify a healthy American diet is worrying, but you could also see it as an opportunity. The future of our diets is a blank slate on which we are free to write our own rules.

TRY TO EAT IN RATIOS, NOT IN ABSOLUTES

Many people today are worried whether they are eating enough protein. Clue: if you are worrying about it, you are probably all right. But as we've seen, it's quite possible that the ratio of protein to carbohydrates in your diet is a little low. You could adjust the ratio by sometimes replacing mashed potatoes or rice with beans or lentils, or by swapping your sandwich for a Danish-style lunch of rye bread topped with smoked fish or a bowl of aromatic curry and a piece of flatbread (half the bread, twice the pleasure).

EAT PROTEIN AND VEGETABLES FIRST AND CARBOHYDRATES LAST

In the West, bread is passed 'round at the start of a meal, but in China, the rice or noodles are traditionally eaten last. There is wisdom in this. A small study published in 2017 involving sixteen people with type 2 diabetes found that leaving carbohydrates to the end of the meal and eating protein and vegetables first lowered blood sugar more effectively than eating the carbohydrates first.[4] An added bonus is that we tend to eat most avidly at the start of the meal, when we are hungriest, so a vegetables-first approach will likely help you to eat more greens. I used this trick with my youngest child when he was picky about vegetables. At the start of every meal, I would serve him a saucer of vegetables to eat with his fingers. Now he is so conditioned to eat vegetables first that he won't touch anything else on his plate until he has eaten every piece of vegetable (unless it's mushrooms, in which case all bets are off). I wouldn't follow this rule obsessively, though. There are many one-pot meals—such as stew and dumplings or Vietnamese pho—where all the elements are joyously combined, and all the better for it.[5]

VARY WHAT YOU EAT

The Global Standard Diet, as we have seen, is based on just a few core items, including animal products, wheat, rice, corn, sugar, refined vegetable oils, and Cavendish bananas. Given that this same diet is causing us ill health on such an epic scale, it seems a good bet that we could improve our health by increasing the variety of what we eat and combining foods in a less average way. This isn't always easy, because what looks like variety often isn't, at the level of ingredients, and many of our choices are predetermined by what is in the shops. But when you get the chance, expand the repertoire of basic foods that you eat. Tim Spector, professor of genetic epidemiology at King's College London, has pointed out that when we eat, we are really feeding not just ourselves but the microbes in our gut, and healthy microbes require a varied diet and benefit from some fermented foods such as pickles or yogurt.[6] Try plums as well as pears; experiment with rye as well as wheat; eat different varieties of apples and unusual cheeses and as many different greens as you can lay your hands on. Don't feel you need to base your diet on ingredients that are "on-trend" or expensive. A herring or a canned sardine can be as good as a piece of wild salmon. If you ever see a grape that isn't seed-less, seize it, and enjoy the novel feeling of the seeds in your mouth.

FIND TIME FOR FOOD

Depending on your patterns of work, there may be no time to cook during your lunch hour or at dinnertime. Finding other times to cook can take a little ingenuity, but it is possible. Any moment you find yourself with a pocket of spare time, consider spending it on the preparation or enjoyment of food. A pressure cooker curry of butter chicken can be quickly cooked one day, refrigerated, and enjoyed the next. Or a few days' worth of lunches can be prepped in advance and kept in the fridge. The new vogue for "meal prepping"—where people plan and cook meals in advance for the week—is one way that the problem of food and time can be squared.

If we never give food the time that it is due, we are effectively say-ing it doesn't matter. I once met a woman who said that people often asked how she had time to cook. "How do you have time to watch television?" she would reply.

LEARN TO COOK THE FOODS THAT YOU WANT YOURSELF TO EAT

So much of the cooking we see on TV or in magazines is virtually useless for everyday life. There are far too many recipes out there for elaborate cakes and dinner party confections and not enough for soups and stews. It's far less valuable to know how to make a perfect silky buttercream frosting than it is to have a trusty formula in your head for a stir-fry or a delicious stew that you can adapt to what's in the fridge.

HAVE UNFASHIONABLE TASTES

Trendy foods are not only overpriced, they are also—as we have seen—more prone to food frauds. You can save money, and make a vote for greater diversity, by choosing overlooked foods. You might choose spring greens instead of kale. Pick blackberries for free in the autumn and freeze them instead of always buying expensive blueberries or other supposed superfoods.

KNOW WHAT YOU ARE EATING

Don't worry too much whether your great-grandmother could recognize what you are eating. It's when you can't recognize the food you have right in front of you that you are in trouble.

One of the most powerful gestures against our globalized food system has been the rebirth of foraging. "Everyone should be a forager," declared the Danish chef Rene Redzepi (of NOMA in Copenhagen) at the World Food Summit in 2017. Redzepi believes in "exploring the edible surroundings" and reclaiming "geography as the foundation of gastronomy." Redzepi can forage leaves from the Danish landscape that are so intensely varied in flavor, they make a mockery of the boring romaine lettuce in the shops. In his cooking, he uses sweet clover that tastes like tonka beans, spicy cresses, and sorrel that is almost lemony in its sourness. The few times I have foraged for wild garlic in the spring, I was staggered that a taste so pungent and green and rich is just there, waiting for us, for free.

What of the rest of us, though? Not everyone has time to go foraging before the morning commute. But we could still bring some of the curiosity of the forager to the food that we eat. We could look at the food in front of us and ask whether it is edible or not. It would be a start if we could see the ingredients on our plate for what they are.

In the spring of 2018, I was in Nanjing with my teenage son. He had been in China for nine months learning Mandarin. We were out with a group of his friends, eating a meal that included a dish of cellophane noodles stir-fried with ground pork and vegetables that he had specially ordered because he said he had eaten it many times and it was delicious. I felt so happy to be there and proud to hear my nearly grown-up son authoritatively ordering food in a foreign language. "I'm glad you chose these mung bean noodles," I said.

"Mum, you're wrong, they are rice noodles," he replied.

We argued to and fro, in an amiable but futile way, about whether cellophane noodles were made from beans or from rice. (They were made from beans, by the way.) Afterward, I thought how strange it was that a person could enjoy a dish so much without actually knowing what it was that he was eating.

We can't escape living and eating in a global market. You can't increase the variety and quality of your diet simply by identifying it. But it helps if you can at least name what's on your plate.

USE YOUR SENSES

One aspect of the current food paradox is that we live in a world of both sensory overload and sensory disconnect. Through advertising and social media, we are bombarded with imagery of food as never before. But the simple process of interacting with food using our senses is often lost. We choose food based on what the label tells us rather than what our senses tell us—a protein bar being a case in point. This is a depressing way to eat, and a slightly inhuman one.

If we overeat, it is partly because except for our sense of taste, our other senses are so underfed that we are wandering around half-starved. When you cook a meal slowly and smell and touch each ingredient as you handle it, you can feel fed by the cooking process itself, and you end up eating less.

Even on busy days, we can feed our senses with food. Keep small pots of herbs in the kitchen or the garden if you have one. When feeling low, pick a mint leaf, rub it on your hand, and inhale deeply.

Try to know your food with your ears, nose, and hands as well as your mouth. Smell it, touch it, and look at it before you taste it. See the way the segments of an orange fall apart. Learn to recognize the difference between fresh and stale garlic, between the acid of lemons and that of vinegar. Try to relish a range of tastes that go beyond sweetness. Appreciate the bitterness of grapefruit and chicory. Notice the sound a really good piece of toast makes when you crunch it. Smell a stick of cinnamon before you add it to a pot of rice. Feel the ridges on a stick of celery. Come to your senses.

ACKNOWLEDGMENTS

THIS BOOK IS THE PRODUCT OF COUNTLESS CONVERSATIONS with many people in different countries who were generous enough to talk to me about the way that they eat. If I have failed to thank some of you by name, I apologize.

My greatest debt is to the many scholars in different fields who have devoted their lives to studying the ways that our eating has changed, and how it has affected human health. Much of the book was inspired by reading the work of Barry Popkin and learning about the "nutrition transition." If you are interested in reading in more detail about the nutrition around the world, I urge you to seek out some of his many, many papers (a small fraction of which are listed in the bibliography).

I benefited from meetings and interviews with numerous experts in different fields. I'd like to thank in particular Hector Abad Faciolince, Lisa Abend, Graeme Arendse, Yemisi Aribisala, Carol Black, Sasha Correa, Prajna Desai, Vikram Doctor, Lynn Dornblaser, Patrick Drake, Chris Elliott, Stuart Flint, Trine Hahnemann, Kerry Hart, Corinna Hawkes, Julian Hearn, Olia Hercules, Fumiaki Imamura, Kathleen Kerridge, Colin Khoury, Michael Krondl, Antoine Lewis, Michael Marmot, Renee McGregor, Theresa Marteau, Chiara Messineo, Jason O'Rourke, Barry Popkin, Rebecca Puhl at the Rudd

Center, Anne Marie Rafferty, Nanna Rögnvaldardóttir, Nilanjana Roy, Alex Rushmer, Joanne Slavin, Zack Szreter, Frank Trentmann, Enrico Vignoli, Dan Wang, Alan Warde, and Eddie Yoon.

Several sections of the book are based on pieces of journalism that first appeared elsewhere. Many thanks to the editors and publications for permission to reproduce these. Some of the section on eating for wellness in Chapter 7 is based on "Why We Fell for Clean Eating," *The Guardian*, August 11, 2017 (with thanks to the editors Clare Longrigg and Jonathan Shainin). Some of my thoughts on meal replacements first appeared in "Food of the Future," *Tank* magazine, Autumn 2016 (with thanks to editor Thomas Roueche). Some of the section on the history of bread in Chapter 3 is based on an essay I wrote for *London Essays*, "No More Daily Bread," June 21, 2016. Some of what I write about Prajna Desai's Mumbai cooking project in Chapter 8 first appeared in *The Observer* in an article called "Social Media and the Great Recipe Explosion," June 18, 2017 (with thanks to editors Gareth Grundy and Allan Jenkins).

My thinking on food has been refined and changed by numerous conversations with friends, colleagues, and family, including Catherine Blyth, Caroline Boileau, Sheila Dillon, Miranda Doyle, Rosalind Dunn, Sophie Hannah, Lucie Johnstone, Ingrid Kopp, Henrietta Lake, Annabel Lee, Ranjita Lohan, Peter McManus, Anne Malcolm, Elfreda Pownall, Sarah Ray, Cathy Runciman, Lisa Runciman, Ruth Runciman, Garry Runciman, Natasha Runciman, Andy Saunders, Abby Scott, Ruth Scurr, Sylvana Tomaselli, Andrew Wilson, and Emma Woolf. Thank you, too, to my dear friends and colleagues at TastEd, including Jason O'Rourke, Ruth Platt, and the teachers and children of St Matthew's primary school.

Because of the breadth of the subject matter, I don't think any book I've written before has had such an interesting journey from concept to final text, and I'm hugely grateful to the guidance and support of my publishers on both sides of the Atlantic for shaping my ideas and words along the way. No one could wish for two more intelligent editors than Lara Heimert at Basic Books and Louise Haines at Fourth Estate. Also at Fourth Estate, I'd like to thank Sarah Thickett and Patrick Hargadon, among others, as well as Steve Gove for his great copy-editing. I'm grateful to the whole team at Basic Books, including Katie Lambright, Kelsey Odorczyk, Liz Wetzel, Issie Ivins, Nancy Sheppard, and Allie Finkel. I can't thank Chin-Yee Lai enough for her superb cover design.

I am tremendously fortunate to have two brilliant agents. For support and guidance of so many kinds, huge thanks to Zoe Pagnamenta in New York and Sarah Ballard at United Agents in London. Also at United Agents, I'd like to thank Eli Keren, for help of so many kinds.

For reading some or all of the manuscript and telling me how to improve it, I'm especially grateful to Eli Keren, Caro Boileau, Tom Runciman, David Runciman, and Emily Wilson.

Needless to say, all the mistakes are mine.

BIBLIOGRAPHY

Abend, Lisa. 2013. "Dan Barber. King of Kale." *TIME*, November 18.

Adair, Linda S., and Barry Popkin. 2012. "Are Child Eating Patterns Being Transformed Globally?" *Obesity Research* 13: 1281–1299.

Adams, Jean, and Martin White. 2015. "Prevalence and Socio-Demographic Correlates of Time Spent Cooking by Adults in the 2005 UK Time Use Survey." *Appetite* 92: 185–191.

Aribisala, Yemisi. 2016. *Longthroat Memoirs: Soups, Sex and Nigerian Taste Buds*. London: Cassava Republic Press.

Aribisala, Yemisi. 2017. *Chimurenga Chronic: We Make Our Own Food*. Self-published, Cape Town, April 2017.

Ascione, Elisa. 2014. "Mamma and the Totemic Robot: Towards an Anthropology of Bimby Food Processors in Italy." In *Food and Material Culture: Proceedings of the Oxford Symposium on Food and Cookery*, edited by Mark McWilliams, 62–69. Devon, UK: Prospect Books.

Bagni, U. V., R. R. Luis, et al. 2013. "Overweight Is Associated with Low Haemoglobin Levels in Adolescent Girls." *Obesity Research and Clinical Practice* 7: e218–e229.

Bahadoran, Zahra, Parvin Mirmiran, et al. 2015. "Fast Food Pattern and Cardiometabolic Disorders: A Review of Current Studies." *Health Promotion Perspectives* 5: 231–240.

Barber, Dan. 2014. *The Third Plate: Field Notes on the Future of Food*. New York: Little, Brown.

Basu, Tanya. 2016. "How Recipe Videos Colonised Your Facebook Feed." *The New Yorker*, May 18.

Becker, Gary. 1965. "A Theory of the Allocation of Time." *Economic Journal* 75: 4935–4917.

Biggs, Joanna. 2013. "Short Cuts." *London Review of Books* 32 (23): 29.

Bloodworth, James. 2018. *Hired: Six Months Undercover in Low-Wage Britain*. London: Atlantic.

Bodzin, Steve. 2014. "Label It: Chile Battles Obesity." *Christian Science Monitor*, January 6.

Bonnell, E. K., C. E. Huggins, et al. 2017. "Influences on Dietary Choices during Day versus Night Shift in Shift Workers: A Mixed Methods Study." *Nutrients* 26: 9.

Boseley, Sarah. 2017. "Amsterdam's Solution to the Obesity Crisis: No Fruit Juice and Enough Sleep." *The Guardian*, April 14.

Bowlby, Rachel. 2000. *Carried Away: The Invention of Modern Shopping*. London: Faber and Faber.

Brannen, Julia, Rebecca O'Connell, and Ann Mooney. 2013. "Families, Meals and Synchronicity: Eating Together in British Dual Earner Families." *Community, Work and Family* 16: 417–434.

Brewis, Alexandra. 2014. "Stigma and the Perpetuation of Obesity." *Social Science Medicine* 118: 152–158.

Brewis, Alexandra A., Amber Wutich, Ashlan Falletta-Cowden, et al. 2011. "Body Norms and Fat Stigma in Global Perspective." *Current Anthropology* 52: 269–276.

Burnett, John. 1983. *Plenty and Want: A Social History of Food in England from 1815 to the Present Day*. Abingdon: Routledge.

Burnett, John. 2004. *England Eats Out: A Social History of Eating Out in England from 1830 to the Present Day*. London: Routledge.

Caballero, Benjamin, and Barry Popkin. 2002. *The Nutrition Transition: Diet and Disease in the Developing World*. Cambridge, MA: Academic Press.

Cahnman, Werner. 1968. "The Stigma of Obesity." *Sociological Quarterly* 9: 283–299.

Cardello, Hank. 2010. *Stuffed: An Insider's Look at Who's (Really) Making America Fat and How the Food Industry Can Fix It*. New York: Ecco.

Caro, Juan Carlos, Shu Wen Ng, Lindsey Smith Taillie, and Barry Popkin. 2017. "Designing a Tax to Discourage Unhealthy Food and Beverage Purchases: The Case of Chile." *Food Policy* 71: 86–100.

Carroll, Abigail. 2013. *Three Squares: The Invention of the American Meal*. New York: Basic Books.

Child, Lydia. 1832. *The Frugal Housewife: Dedicated to Those Who Are Not Ashamed of Economy*. London: T. T. and J. Tegg.

Choi, S. K., H. J. Choi, et al. 2008. "Snacking Behaviours of Middle and High School Students in Seoul." *Korean Journal of Community Nutrition* 13: 199–206.

Clements, Kenneth W., and Dongling Chen. 2010. "Affluence and Food: A Simple Way to Infer Incomes." *American Journal of Agricultural Economics* 92 (4): 909–926.

Clifton, Peter, Sharayah Carter, et al. 2015. "Low Carbohydrate and Ketogenic Diets in Type 2 Diabetes." *Current Opinion in Lipidology*, 26: 594–595.

Colchero, M. A., Marina Molina, et al. 2017. "After Mexico Implemented a Tax, Purchases of Sugar-Sweetened Beverages Decreased and Water Increased." *Journal of Nutrition*, 147: 1552–1557.

Cooper, Derek. 2000. *Snail Eggs and Samphire: Dispatches from the Food Front*. London: Macmillan.

Coudray, Guillaume. 2017. *Cochonneries: Comment la charcuterie est devenue un poison*. Paris: La Découverte.

Cowen, Tyler. 2012. *An Economist Gets Lunch: New Rules for Everyday Foodies*. New York: Plume.

Cuadra, Cruz Miguel Ortiz. 2006. *Eating Puerto Rico: A History of Food, Culture, and Identity*. Translated by Russ Davidson. Chapel Hill: University of North California Press.

Currie, Janet, Stefano DellaVigna, Enrico Mofretti, et al. 2010. "The Effect of Fast Food Restaurants on Obesity and Weight Gain." *American Economic Journal* 2: 32–63.

Datamonitor. 2015. "Savoury Snack Industry Profile USA." November.

David, Elizabeth. 2010. *Spices, Salt and Aromatics in the English Kitchen*. London: Grub Street.

De Crescenzo, Sarah. 2017. "Perfect Bar Finds Missing Ingredient." *San Diego Business Journal*, July 27.

Demmler, Kathrin, Olivier Ecker, et al. 2018. "Supermarket Shopping and Nutritional Outcomes: A Panel Data Analysis for Urban Kenya." *World Development* 102: 292–303.

Desai, Prajna. 2015. *The Indecisive Chicken: Stories and Recipes from Eight Dharavi Cooks*. Mumbai: SNEHA.

De Vries, Gerard, Josta de Hoog, et al. 2016. *Towards a Food Policy*. The Hague: Netherlands Scientific Council.

DiMeglio, D. P., and R. D. Mattes. 2000. "Liquid versus Solid Carbohydrate: Effects on Food Intake and Body Weight." *International Journal of Obesity Related Metabolic Disorders* 24: 794–800.

Doak, Colleen, Linda Adair, Carlos Monteiro, and Barry Popkin. 2000. "Overweight and Underweight Coexist within Households in Brazil, China and Russia." *Journal of Nutrition* 130: 2965–2971.

Dunn, Elizabeth. 2018. "How Delivery Apps May Put Your Favorite Restaurant Out of Business." *The New Yorker*, February 3.

Dunn, Rob. 2017. *Never Out of Season: How Having the Food We Want When We Want It Threatens Our Food Supply and Our Future.* New York: Little, Brown and Company.

Eckhardt, Cara. 2006. "Micronutrient Malnutrition, Obesity and Chronic Disease in Countries Undergoing the Nutrition Transition: Potential Links and Programme/Policy Implications." International Food Policy Research Institute, FCND discussion papers.

Erikson, Gary. 2004. *Raising the Bar: Integrity and Passion in Life and Business; The Story of Clif Bar Inc.* New York: Jossey-Bass.

Fahey, Jed, and Eleanore Alexander. 2015. "Opinion: Current Fruit Breeding Practices: Fruitful or Futile?" October 19. Accessed October 2018. https://www.freshfruitportal.com/news/2015/10/19/opinion-current-fruit-breeding-practices-fruitful-or-futile.

Fiolet, Thibault, Bernard Srour, et al. 2018. "Consumption of Ultra-Processed Foods and Cancer Risk: Results from Nutrinet-Santé Prospective Cohort." *British Medical Journal* 360: k322.

Fisher, J. O., G. Wright, A. N. Herman, et al. 2015. "'Snacks Are Not Food': Low-Income, Urban Mothers' Perceptions of Feeding Snacks to Their Pre-School Children." *Appetite* 84: 61–67.

Food and Agriculture Organization of the United Nations. 2016. "Table and Dried Grapes: FAO-OIV Focus 2016." FAO and OIV. Accessed October 2018. http://www.fao.org/3/a-i7042e.pdf.

Fresco, Louise. 2015. *Hamburgers in Paradise: The Stories behind the Food We Eat.* Princeton, NJ: Princeton University Press.

Fu, Wenge, Vasant P. Gandhi, Lijuan Cao, Hongbo Liu, and Zhangyue Zhou. 2012. "Rising Consumption of Animal Products in China and India: National and Global Implications." *China and World Economy*, 20: 88–106.

Fulkerson, J. A., N. Larson, et al. 2014. "A Review of Associations between Family or Shared Meal Frequency and Dietary and Weight Status

Outcomes across the Lifespan." *Journal of Nutrition Education and Behavior* 46: 2–19.

Gakidou, Emmanuela. 2017. "Global, Regional, and National Comparative Risk Assessment of 84 Behavioural, Environmental and Occupational, and Metabolic Risks or Clusters of Risks, 1990–2016: A Systematic Analysis for the Global Burden of Disease Study 2016." *Lancet* 390: 1345–422. Accessed September 2018. https://www.thelancet.com/pdfs /journals/lancet/PIIS0140-6736(17)32366-8.pdf.

Guthrie, J. F. 2002. "Role of Food Prepared Away from Home in the American Diet, 1977–8 versus 1994–6: Changes and Consequences." *Journal of Nutrition Education and Behaviour* 34: 140–150.

Haddad, Lawrence, Corinna Hawkes, Patrick Webb, et al. 2016. "A New Global Research Agenda for Food." *Nature*, November 30: 30–32.

Haggblade, S., K. G. Duodu, et al. 2016. "Emerging Early Actions to Bend the Curve in Sub-Saharan Africa's Nutrition Transition." *Food Nutrition Bulletin* 37: 219–241.

Hahnemann, Trine. 2016. *Scandinavian Comfort Food: Embracing the Art of Hygge*. London: Quadrille.

Hamilton, Lisa. 2014. "The Quinoa Quarrel: Who Owns the World's Greatest Superfood?" *Harper's Magazine*. May. Accessed October 2018. https:// harpers.org/archive/2014/05/the-quinoa-quarrel/.

Hansen, Henning O. 2013. *Food Economics: Industry and Markets*. Abingdon: Routledge.

Harvey, Simon. 2017. "Strong UK Performance Boosts Arla Foods' Figures." *Just-Food Global News*, August 27.

Hawkes, Corinna. 2004. "The Role of Foreign Direct Investment in the Nutrition Transition." *Public Health Nutrition* 8: 357–365.

Hawkes, Corinna. 2006. "Uneven Dietary Development: Linking the Policies and Processes of Globalization with the Nutrition Transition, Obesity, and Diet-Related Chronic Diseases." *Globalization and Health* 2: 4.

Hawkes, Corinna. 2012. "Food Policies for Healthy Populations and Healthy Economies." *British Medical Journal* 344: 27–29.

Hawkes, Corinna, T. G. Smith, J. Jewell, et al. 2015. "Smart Policies for Obesity Prevention." *Lancet* 385: 2410–2421.

Hawkes, Corinna, Sharon Friel, Tim Lobstein, and Tim Lang. 2012. "Linking Agricultural Policies with Obesity and Noncommunicable Diseases: A New Perspective for a Globalizing World." *Food Policy* 37: 343–353.

Hercules, Olia. 2015. *Mamushka: Recipes from Ukraine and Beyond*. London: Mitchell Beazley.

Hess, Amanda. 2017. "The Hand Has Its Social Media Moment." *New York Times*, October 11.

Hess, Julie, and Joanne Slavin. 2014. "Snacking for a Cause: Nutritional Insufficiencies and Excesses of U.S. Children, a Critical Review of Food Consumption Patterns and Macronutrient and Micronutrient Intake of U.S. Children." *Nutrients* 6: 4750–4759.

Hollands, Gareth J., Ian Shemilt, Theresa Marteau, et al. 2013. "Altering Micro-Environments to Change Population Health Behaviour: Towards an Evidence Base for Choice Architecture." *British Medical Council Public Health* 12: 1218.

Hong, E. 2016. "Why Some Koreans Make $10,000 a Month to Eat on Camera." https://qz.com/592710/why-some-koreans-make-10000-a-month-to -eat-on-camera/.

Hu, Winnie. 2016. "With Food Hub, Premium Produce May Reach More New Yorkers' Plates." *New York Times*, September 5.

Imamura, Fumiaki, Renata Micha, Shahab Khatibzadeh, et al. 2015. "Dietary Quality among Men and Women in 187 countries in 1990 and 2010: A Systematic Assessment." *Lancet Global Health* 3: e132–42.

Imamura, Fumiaki, Laura O'Connor, Zheng Ye, et al. 2015. "Consumption of Sugar-Sweetened Beverages, Artificially Sweetened Beverages and Fruit Juice and Incidence of Type 2 Diabetes." *British Medical Journal* 351: h3576.

Jabs, J., and C. M. Devine. 2006. "Time Scarcity and Food Choices: An Overview." *Appetite* 47: 196–204.

Jacobs, Andrew. 2018. "In Sweeping War on Obesity, Chile Slays Tony the Tiger." *New York Times*, February 7. Accessed October 2018. https:// www.nytimes.com/2018/02/07/health/obesity-chile-sugar-regulations .html.

Jacobs, Andrew, and Matt Richtel. 2017. "How Big Business Got Brazil Hooked on Junk Food." *New York Times*, September 16.

Jacobs, Marc, and Peter Scholliers, eds. 2003. *Eating Out in Europe: Picnics, Gourmet Dining and Snacks since the Late Eighteenth Century*. London: Berg.

Jacobsen, Sven-Erik. 2011. "The Situation for Quinoa and Its Production in Southern Bolivia: From Economic Success to Environmental Disaster." *Journal of Agronomy and Crop Science*, May 22.

Jahns, Lisa, Anna Maria Siega-Riz, and Barry Popkin. 2001. "The Increasing Prevalence of Snacking among US Children from 1977 to 1996." *Journal of Pediatrics* 138: 493–498.

Jastran, Margaret, Carole Bisogni, et al. 2009. "Eating Routines: Embedded, Value Based, Modifiable and Reflective." *Appetite* 52: 127–136.

Johansen, Signe. 2018. *Solo: The Joy of Cooking for One*. London: Bluebird.

Kammlade, Sarah, and Colin Khoury. 2017. "Five Surprising Ways People's Diets Have Changed over the Past Fifty Years." International Center for Tropical Agriculture (CIAT), a CGIAR Research Center. https://blog.ciat.cgiar.org /five-surprising-ways-peoples-diets-have-changed-over-the-past-50-years/.

Kamp, David. 2006. *The United States of Arugula: The Sun Dried, Cold Pressed, Dark Roasted, Extra Virgin Story of the American Food Revolution*. New York: Broadway Books.

Kant, Ashima, and Barry I. Graubard. 2004. "Eating Out in America 1987–2000: Trends and Nutritional Correlates." *Preventive Medicine* 38: 243–249.

Kant, Ashima, and Barry Graubard. 2015. "40 Year Trends in Meal and Snack Eating Behaviors of American Adults." *Journal of the Academy of Nutrition and Dietetics*, 2212–2672.

Kateman, Brian, ed. 2017. *The Reducetarian Solution*. New York: Tarcher/ Putnam.

Kearney, John. 2010. "Food Consumption Trends and Drivers." *Philosophical Transactions of the Royal Society of London B* 365 (September 27): 2793–2807.

Keats, Sharada, and Steve Wiggins. 2014. *Future Diets: Implications for Agriculture and Food Prices*. London: Overseas Development Institute.

Kelly, Bridget, Jason C. G. Halford, Emma J. Boyland, et al. 2010. "Television Food Advertising to Children: A Global Perspective." *American Journal of Public Health* 100 (9): 1730–1736.

Khaleeli, Homa. 2016. "The Truth about Working for Deliveroo, Uber and the On-Demand Economy." *The Guardian*, June 15.

Khoury, Colin. 2017. "How Diverse Is the Global Diet?" International Center for Tropical Agriculture (CIAT), a CGIAR Research Center. http://blog .ciat.cgiar.org/how-diverse-is-the-global-diet/.

Khoury, Colin, Anne D. Bjorkman, et al. 2014. "Increasing Homogeneity in Global Food Supplies and the Implications for Food Security." *Proceedings of the National Academy of Sciences of the United States of America* 111: 4001–4006.

Khoury, Colin, Harold A. Achicanoy, et al. 2016. "Origins of Food Crops Connects Countries Worldwide." *Proceedings of the Royal Society B* 283.

Kim, Soowon, Soojae Moon, and Barry Popkin. 2000. "The Nutrition Transition in South Korea." *American Journal of Clinical Nutrition* 71: 44–53.

Kludt, Amanda, and Daniel Geneen. 2018. "Dan Barber Wants to Revolutionize the Way the World Grows Vegetables." Eater.com, March 1.

Konnikova, Maria. 2018. *The Confidence Game: The Psychology of the Con and Why We Fall for It Every Time.* London: Canongate Books.

Krishnan, Supriya, Patricia F. Coogan, et al. 2010. "Consumption of Restaurant Foods and Incidence of Type 2 Diabetes in African American Women." *American Journal of Clinical Nutrition* 91: 465–471.

Kvidahl, Melissa. 2017. "Market Trends: Bars." *Snack Food and Wholesale Bakery* 106: 14–20.

Lang, Tim, and Pamela Mason. 2017. *Sustainable Diets: How Ecological Nutrition Can Transform Consumption and the Food System.* Abingdon: Routledge.

Lang, Tim, and Erik Millstone. 2008. *The Atlas of Food: Who Eats What, Where, and Why.* London: Routledge.

Laudan, Rachel. 2016. "'A Good Cook': On My Mother's Hundredth Birthday." www.rachellaudan.com, October 12.

Lawrence, Felicity. 2004. *Not on the Label: What Really Goes into the Food on Your Plate.* London: Penguin.

La Vecchia, Carlo, and Luis Serra Majem. 2015. "Evaluating Trends in Global Dietary Patterns." *Lancet Global Health* 3: e114–e115.

Lawler, Andrew. 2016. *How the Chicken Crossed the World: The Story of the Bird That Powers Civilisations.* London: Gerald Duckworth.

Lee, H. S., K. J. Duffey, and Barry Popkin. 2012. "South Korea's Entry to the Global Food Economy: Shifts in Consumption of Food between 1998 and 2009." *Asia Pacific Journal of Clinical Nutrition* 21: 618–629.

Lee, Min-June, Barry Popkin, and Soowon Kim. 2002. "The Unique Aspects of the Nutrition Transition in South Korea: The Retention of Healthful Elements in Their Traditional Diet." *Public Health Nutrition* 5: 197–203.

Levy-Costa, Renata, et al. 2005. "Household Food Availability in Brazil: Distribution and Trends (1974–2003)." *Rev. Saúde Pública* 39 (4): 530–540.

Ley, Sylvia H., An Pan, et al. 2016. "Changes in Overall Diet Quality and Subsequent Type 2 Diabetes Risk: Three U.S. Prospective Cohorts." *Diabetes Care.* http://care.diabetesjournals.org/content/39/11/2011.

Lloyd, Susan. 2014. "Rose Vouchers for Fruit and Veg—An Evaluation Report." City University, London. www.alexandrarose.org.uk.

Lopez, Oscar, and Andrew Jacobs. 2018. "In a Town with Little Water, Coca-Cola Is Everywhere. So Is Diabetes." *New York Times*, July 14. Accessed October 2018. https://www.nytimes.com/2018/07/14/world/americas/mexico-coca -cola-diabetes.html.

Lymbery, Philip, with Isabel Oakshott. 2014. *Farmageddon: The True Cost of Cheap Meat*. London: Bloomsbury.

McGregor, Renee. 2017. *Orthorexia: When Healthy Eating Goes Bad*. London: Nourish Books.

Manjoo, Farhad. 2017. "How Buzzfeed's Tasty Conquered Online Food." *New York Times*, July 27.

Markley, Klare S. 1951. *Soybeans and Soybean Products*. New York: Interscience Publishers.

Marmot, Michael, and S. L. Syme. 1976. "Acculturations and Coronary Heart Disease in Japanese-Americans." *American Journal of Epidemiology* 104: 225–247.

Mattes, R. D. 2006. "Fluid Energy—Where's the Problem?" *Journal of the American Dietietic Association* 106: 1956–1961.

Maumbe, Blessing. 2012. "The Rise of South Africa's Quick Service Restaurant Industry." *Journal of Agribusiness in Developing and Emerging Economies* 2: 147–166.

Mead, Rebecca. 2013. "Just Add Sugar." *The New Yorker*, November 4.

Meades, Jonathan. 2014. *An Encyclopedia of Myself*. London: Fourth Estate.

Mellentin, Julian. 2018. "Keeping Trend Connecting: Both Siggi's and Noosa Have Been Successful in the US by Leveraging Key Trends in Dairy." *Dairy Industries International* 83: 14.

Mendis, Shanti, et al. 2014. "Global Status Report on Non-Communicable Diseases." World Health Organization.

Menzel, Peter, and Faith d'Aluiso. 2005. *Hungry Planet: What the World Eats*. New York: Ten Speed Press.

Micha, Renata, Shahab Khatibzadeh, Peilin Shi, et al. 2015. "Global, Regional and National Consumption of Major Food Groups in 1990 and 2010: A Systematic Analysis Including 266 Country-Specific Nutrition Surveys Worldwide." *British Medical Journal Open* 5 (9). https://bmjopen.bmj .com/content/5/9/e008705.

Micha, Renata, and Dariush Mozaffarian. 2010. "Saturated Fat and Cardio-metabolic Risk Factors: Coronary Heart Disease, Stroke and Diabetes: A Fresh Look at the Evidence." *Lipids* 45 (10): 893–905.

Micha, Renata, Masha L. Shulkin, et al. 2017. "Etiologic Effects and Optimal Intakes of Foods and Nutrients for Risk of Cardiovascular Diseases and Diabetes: Systematic Reviews and Meta-Analyses from the Nutrition and Chronic Diseases Expert Group (NutriCoDE)." *Public Library of Science*, April 27.

Mikkila, V., H. Vepsalainen, et al. 2015. "An International Comparison of Dietary Patterns in 9–11-year-old children." *International Journal of Obesity Supplements* 5: S17–S21.

Millstone, Erik, and Tim Lang. 2008. *The Atlas of Food: Who Eats What, Where and Why*. 2nd ed. London: Earthscan.

Mintel. 1985. *Crisps, Nuts and Savoury Snacks*. London: Mintel Publications.

Monteiro, Carlos. 2009. "Nutrition and Health: The Issue Is Not Food, nor Nutrients, So Much as Processing." *Public Health Nutrition* 12: 729–731.

Monteiro, Carlos, Geoffrey Cannon, et al. 2016. "NOVA: The Star Shines Bright." *World Nutrition* 7, nos. 1–3 (January–March). Accessed July 2018. http://archive.wphna.org/wp-content/uploads/2016/01/WN-2016 -7-1-3-28-38-Monteiro-Cannon-Levy-et-al-NOVA.pdf.

Monteiro, C. A., J.-C Moubarac, G. Cannon, S. W. Ng, and Barry Popkin. 2013. "Ultra-Processed Products Are Becoming Dominant in the Global Food System." *Obesity Reviews* 14: 21–28.

Morley, Katie. 2016. "Smoothie Craze Sees Berry Sales Reach £1bn— Overtaking Apples and Bananas." *Daily Telegraph*, 23 May.

Moss, Michael. 2014. *Salt, Sugar, Fat: How the Food Giants Hooked Us*. London: W. H. Allen.

Murray, Christopher, et al. 2016. "Global, Regional, and National Comparative Risk Assessment of 79 Behavioural, Environmental and Occupational, and Metabolic Risks or Clusters of Risks, 1990–2015: A Systematic Analysis for the Global Burden of Disease Study 2015." *Lancet* 388: 1639–1724.

Nago, Eunice, Carl Lachat, et al. 2010. "Food, Energy and Macronutrient Contribution of Out-of-Home Foods in School-Going Adolescents in Cotonou, Benin." *British Journal of Nutrition* 103: 281–288.

Nestle, Marion. 2018. *Unsavory Truth: How Food Companies Skew the Science of What We Eat*. New York: Basic Books.

Nguyen, Binh, and Lisa Powell. 2014. "The Impact of Restaurant Consumption among US Adults: Effects on Energy and Nutrient Intakes." *Public Health Nutrition* 17: 2445–2452.

Nielsen, Samara Joy, and Barry Popkin. 2004. "Changes in Beverage Intake between 1977 and 2001." *American Journal of Preventive Medicine* 27: 205–210.

Norberg, Johan. 2016. *Progress: Ten Reasons to Look Forward to the Future.* London: Oneworld Publications.

"Now Comes Quinoa: It's a Substitute for Spinach, Dear Children All." 1954. *New York Times*, March 7.

O'Brien, Charmaine. 2013. *The Penguin Food Guide to India.* London: Penguin.

Oliver, Brian. 2016. "Welcome to Skyr, the Viking 'Superfood' Waking Up Britain." *Observer*, November 27.

Olson, Parmy. 2016. "Here's How Deliveroo Built an Army of 5000 Drivers in Just 3 Years." *Forbes*, February 17.

Orfanos, P., et al. 2007. "Eating Out of Home and Its Correlates in 10 European Countries." *Public Health and Nutrition* 10: 1515–1525.

Packer, Robert. 2013. "Pomegranate Juice Adulteration." *Food Safety Magazine*, February, online edition.

Perelman, Deb. 2018. "Never Cook at Home." *New York Times*, August 25.

Piernas, Carmen, and Barry Popkin. 2009. "Snacking Increased among U.S. Adults between 1977 and 2006." *Journal of Nutrition* 140 (2): 325–332.

Piernas, Carmen, and Barry Popkin. 2010. "Trends in Snacking among U.S. Children." *Health Affairs* 29: 398–404.

Pomiane, Edouard de. 2008. *Cooking in Ten Minutes; or, The Adaptation to the Rhythm of Our Time.* Translated by Peggie Benton. London: Serif.

Popkin, Barry. 2001. "The Nutrition Transition and Obesity in the Developing World." *Journal of Nutrition* 131: 871S–873S.

Popkin, Barry. 2002. "The Dynamics of the Dietary Transition in the Developing World." In *The Nutrition Transition: Diet and Disease in the Developing World*, edited by Benjamin Caballero and Barry Popkin, 111–128. Cambridge, MA: Academic Press.

Popkin, Barry. 2009. *The World Is Fat: The Fads, Trends, Policies, and Products That Are Fattening the Human Race.* New York: Avery.

Popkin, Barry. 2011. "Contemporary Nutrition Transition: Determinants of Diet and Its Impact on Body Composition." *Proceedings of the Nutrition Society* 70: 82–91.

Popkin, Barry, Linda Adair, and Shu Wen Ng. 2012. "Now and Then: The Global Nutrition Transition: The Pandemic of Obesity in Developing Countries." *Nutrition Review* 70: 3–21.

Popkin, Barry, Shufa du Fengying, and Zhai Bing Zhang. 2010. "Cohort Profile: The China Health and Nutrition Survey—Monitoring and Understanding Socio-Economic and Health Change in China, 1989–2011." *International Journal of Epidemiology* 39: 1435–1440.

Popkin, Barry, and P. Gordon-Larsen. 2004. "The Nutrition Transition: Worldwide Obesity Dynamics and Their Determinants." *International Journal of Obesity* 28: S2–S9.

Popkin, Barry, and Corinna Hawkes. 2016. "Sweetening of the Global Diet, Particularly Beverages: Patterns, Trends and Policy Responses." *Lancet Diabetes Endocrinology* 4: 174–186.

Pollan, Michael. 2013. *Cooked: A Natural History of Transformation*. New York: Penguin Books.

Powell, L. M., and Y. Bao. 2009. "Food Prices, Access to Food Outlets and Child Weight." *Economics and Human Biology* 7: 64–72.

Puhl, Rebecca, and Chelsea Heuer. 2010. "Obesity Stigma: Important Considerations for Public Health." *American Journal of Public Health* 100: 1019–1028.

Rao, Tejal. 2018. "Seeds Only a Plant Breeder Could Love, until Now." *New York Times*, February 27.

Roberts, Paul. 2008. *The End of Food*. New York: Houghton, Mifflin Harcourt.

Rögnvaldardóttir, Nanna. 2002. *Icelandic Food and Cookery*. New York: Hippocrene Books.

Rogers, Anthony, Alistair Woodward, Boyd Swinburn, and William Dietz. 2018. "Prevalence Trends Tell Us What Did Not Precipitate the US Obesity Epidemic." *Lancet* 3, no. 4 (April): 162–163. https://www.thelancet.com/journals/lanpub/article/PIIS2468-2667(18)30021-5/fulltext.

Richardson, S. A., N. Goodman, et al. 1961. "Cultural Uniformity in Reaction to Physical Disabilities." *American Sociological Review* 26: 241–247.

Rickertsen, Kyrre, and Wen S. Chern. 2003. *Health, Nutrition and Food Demand*. Wallingford: Cabi International.

Robinson, John, and Geoffrey Godbey. 1997. *Time for Life: The Surprising Ways Americans Use Their Time*. Philadelphia: Penn State University Press.

Saladino, Dan. 2017. "Hunting with the Hadza." BBC Radio 4 *Food Programme*. First broadcast July 2.

Sax, David. 2014. *The Tastemakers: Why We're Crazy for Cupcakes but Fed Up with Fondue*. New York: Public Affairs.

Schmit, Todd M., and Harry M. Kaiser. 2003. "The Impact of Dietary Cholesterol Concerns on Consumer Demand for Eggs in the USA." In *Health,*

Nutrition and Food Demand, edited by Wen S. Chern and Kyrre Rickertsen, 203–222. Wallingford: Cabi International.

Schwartz, Barry. 2004. *The Paradox of Choice: Why Less Is More*. New York: Harper Perennial.

Seccia, Antonio, Fabio G. Santeramo, and Gianluca Nardone. 2015. "Trade Competitiveness in Table Grapes: A Global View." *Outlook on Agriculture* 44: 127–134.

Severson, Kim. 2016. "The Dark (and Often Dubious) Art of Forecasting Food Trends." *New York Times*, December 27.

"Should We Officially Recognise Obesity as a Disease?" 2017. Editorial, *Lancet Diabetes and Endocrinology* 5 (June 7).

Short, Frances. 2006. *Kitchen Secrets: The Meaning of Cooking in Everyday Life*. London: Berg.

Shukla, Alpana P., Jeselin Andono, et al. 2017. "Carbohydrate-Last Meal Pattern Lowers Postprandial Glucose and Insulin Excursions in Type 2 Diabetes." *BMJ Open Diabetes Research Care* 5 (1): 1–5.

Smil, Vaclav. 2002. "Food Production." In *The Nutrition Transition: Diet and Disease in the Developing World*, edited by Benjamin Caballero and Barry Popkin, 25–50. Cambridge, MA: Academic Press.

Smith, L. P., S. W. Ng, and B. M. Popkin. 2013. "Trends in U.S. Home Food Preparation and Consumption. Analysis of National Nutrition Surveys and Time Use Studies from 1965–6 to 2007–8." *Nutrition Journal* 12: 45.

Sole-Smith, Virginia. 2018. *The Eating Instinct: Food Culture, Body Image and Guilt in America*. New York: Henry Holt and Co.

Soskin, Anthony B. 1988. *Non-traditional Agriculture and Economic Development: The Brazilian Soybean Expansion 1962–1982*. Westport, CT: Praeger.

Spector, Tim. 2015. *The Diet Myth: The Real Science behind What We Eat*. London: Weidenfeld and Nicolson.

Spector, Tim. 2017. "I Spent Three Days as a Hunter-Gatherer to See If It Would Improve My Gut Health." *The Conversation*, June 30. https://theconversation.com/i-spent-three-days-as-a-hunter-gatherer-to-see-if-it-would-improve-my-gut-health-78773.

Steel, Carolyn. 2013. *Hungry City: How Food Shapes Our Lives*. Reissue. London: Vintage.

Tandoh, Ruby. 2018. *Eat Up: Food, Appetite and Eating What You Want*. London: Serpent's Tail.

Thornhill, Ted. 2014. "Crumbs, Would You Pay £2.40 for a Slice of TOAST? New 'Artisanal Toast Bars' Springing Up in San Francisco Selling Posh Grilled Bread." *Daily Mail*, January 25. Accessed October 2018. https://www.dailymail.co.uk/news/article-2545832/The-new-artisanal-toast-bars-springing-San-Francisco-UK.html.

Tomiyama, A. Janet. 2014. "Weight Stigma Is Stressful: A Review of Evidence for the Cyclic Obesity/Weight-Based Stigma Model." *Appetite* 82: 8–15.

Townsend, N. 2015. "Shorter Lunch Breaks Lead Secondary-School Students to Make Less Healthy Dietary Choices: Multilevel Analysis of Cross-Sectional National Survey Data." *Public Health Nutrition* 18: 1626–1634.

Trentmann, Frank. 2016. *Empire of Things: How We Became a World of Consumers, from the Fifteenth Century to the Twenty-First*. London: Penguin.

Tshukudu, Mpho, and Anna Trapido. 2016. *Eat Ting: Lose Weight. Gain Health. Find Yourself.* Cape Town: Quivertree Publications.

Van Dam, Rob M., and David Hunter. 2012. "Biochemical Indicators of Dietary Intake." In *Nutritional Epidemiology*, edited by Walter Willett, 150–213. Oxford: Oxford Scholarship Online.

Van den Bos, Lianne. 2015. "The War of 'Origin' Yoghurts." *Food Magazine*, July 28.

Vorster, Hester H., Annamarie Kruger, et al. 2011. "The Nutrition Transition in Africa: Can It Be Steered in a More Positive Direction?" *Nutrients* 3: 429–441.

Walvin, James. 2018. *Sugar: The World Corrupted, from Obesity to Slavery*. New York: Pegasus.

Wang, Dantong, Klazine van der Horst, et al. 2018. "Snacking Patterns in Children: A Comparison between Australia, China, Mexico and the U.S." *Nutrients* 10: 198.

Wang, Dong D., Cindy Leung, et al. 2014. "Trends in Dietary Quality among Adults in the United States 1999 through 2010." *JAMA International Medicine* 174: 1587–1595.

Wang, Zhihong, Fengying Zhai, Bing Zhang, and Barry Popkin. 2012. "Trends in Chinese Snacking Behaviors and Patterns and the Social-Demographic Role between 1991 and 2009." *Asia Pacific Journal of Clinical Nutrition* 21: 253–262.

Warde, Alan. 2016. *The Practice of Eating*. Cambridge: Polity.

Warde, Alan, and Lydia Martens. 2009. *Eating Out: Social Differentiation, Consumption and Pleasure*. Cambridge: Cambridge University Press.

Warde, Alan, and Luke Yates. 2015. "The Evolving Content of Meals in Great Britain: Results of a Survey in 2012 in Comparison with the 1950s." *Appetite* 84: 299–308.

Warde, Alan, and Luke Yates. 2016. "Understanding Eating Events: Snacks and Meal Patterns in Great Britain." *Food, Culture & Society* 20 (1): 15–35.

Warren, Geoffrey. 1958. *The Foods We Eat.* London: Cassell.

Watt, Abigail. 2015. "India's Confectionery Market to Grow 71% in 4 Years." *Candy Industry* 180 (3): 12–13.

Whitney, Alyse. 2017. "Honeynut Squash Is a Tiny Squash with a Big History." *Bon Appétit*, November 30.

Whittle, Natalie. 2016. "The Fight against Food Fraud." *Financial Times*, March 24.

Widdecombe, Lizzie. 2014. "The End of Food." *The New Yorker*, May 5.

Wiggins, Steve, and Sharada Keats. 2015. *The Rising Cost of a Healthy Diet: Changing Relative Prices of Foods in High-Income and Emerging Economies.* London: Overseas Development Institute.

Willett, Walter. 2013. *Nutritional Epidemiology.* 3rd ed. New York: Oxford University Press.

Wilson, Bee. 2008. *Swindled: From Poison Sweets to Counterfeit Coffee.* London: John Murray.

Winson, Anthony. 2013. *The Industrial Diet: The Degradation of Food and the Struggle for Healthy Eating.* Vancouver: UBC Press.

Wolf, A., G. A. Bray, and B. M. Popkin. 2008. "A Short History of Beverages and How Our Body Treats Them." *Obesity Reviews* 9: 151–164.

Yano, K., W. C. Blackwelder, et al. 1979. "Childhood Cultural Experience and the Incidence of Coronary Heart Disease in Hawaii Japanese Men." *American Journal of Epidemiology* 109: 440–450.

Yajnik, C. S. 2018. "Confessions of a Thin-Fat Indian." *European Journal of Clinical Nutrition* 72: 469–473.

Yajnik, C. S., and John S. Yudkin. 2004. "The Y-Y Paradox." *Lancet* 363: 163.

Yajnik, C. S., C. H. D. Fall, and K. A. Coyaji. 2003. "Neonatal Anthropometry: The Thin-Fat Indian Baby. The Pune Maternal Nutrition Study." *International Journal of Obesity* 27: 173–180.

Yoon, Eddie. 2017. "The Grocery Industry Confronts a New Problem: Only 10% of Americans Love Cooking." *Harvard Business Review*, September 27.

Zaraska, Marta. 2016. *Meathooked: The History and Science of Our 2.5-Million-Year Obsession with Meat.* New York: Basic Books.

Zaraska, Marta. 2017. "Bitter Truth: How We're Making Fruit and Veg Less Healthy." *New Scientist*, September 2.

Zhai, F. Y., S. F. Du, et al. 2014. "Dynamics of the Chinese Diet and the Role of Urbanicity, 1991–2011." *Obesity Reviews* 1: 16–26.

Zhou, Yijing, Shufa Du, et al. 2015. "The Food Retails Revolution in China and Its Association with Diet and Health." *Food Policy* 55: 92–100.

Zupan, Z., A. Evans, D.-L. Couturier, and Theresa Marteau. 2017. "Wine Glass Size in England from 1700 to 2017: A Measure of Our Time." *British Medical Journal* 359. https://www.bmj.com/content/359/bmj.j5623.

NOTES

INTRODUCTION: THE GATHERERS AND THE HUNTED

1 Fahey and Alexander 2015; Zaraska 2017.
2 Food and Agriculture Organization of the United Nations 2016; Seccia, Santeramo, and Nardone 2015.
3 Murray et al. 2016; Gakidou 2017.
4 Micha, Khatibzadeh, et al. 2015.
5 Lang and Mason 2017.
6 Cited in Rogers, Woodward, et al. 2018.
7 Rogers, Woodward, et al. 2018; Jacobs and Richtel 2017.
8 Cardello 2010, 19; Jacobs and Richtel 2017.
9 Cardello 2010, 10.
10 Moss 2014.
11 Jahns, Siega-Riz, et al. 2001.
12 Sarah Boseley, 2018, "'The Mediterranean Diet Is Gone': Region's Children Are Fattest in Europe," *The Guardian*, May 24, https://www .theguardian.com/society/2018/may/24/the-mediterranean-diet-is-gone -regions-children-are-fattest-in-europe.
13 Konnikova 2018.
14 Warde and Yates 2015.
15 Gakidou 2017; Micha, Shulkin, et al. 2017.
16 Simpson and Raubenheimer 2012.
17 Simpson and Raubenheimer 2012.

CHAPTER ONE: THE FOOD TRANSITION

1   Caballero and Popkin 2002.
2   Nicholas Kristof, 2017, "Why 2017 May Be the Best Year Ever," *New York Times*, January 21, https://www.nytimes.com/2017/01/21/opinion/sunday/why-2017-may-be-the-best-year-ever.html?_r=0.
3   Norberg 2016.
4   Smil 2002.
5   Norberg 2016.
6   Nick Squires, 2006, "Overweight People Now Outnumber the Hungry," *The Telegraph*, August 15, http://www.telegraph.co.uk/news/uknews/1526403/Overweight-people-now-outnumber-the-hungry.html.
7   Haddad, Hawkes, et al. 2016; Micha, Khatibzadeh, et al. 2015.
8   Ley, Pan, et al. 2016; Imamura, O'Connor, et al. 2015; Popkin 2009, 17.
9   "Diabetes UK comments on the Rise in Type 2 Diabetes in Children," Diabetes UK, August 12, 2017, accessed November 2018, https://www.diabetes.org.uk/about_us/news/type-2-diabetes-in-children.
10   Imamura, Micha, et al. 2015.
11   Micha and Mozaffarian 2010.
12   Willett 2013; Van Dam and Hunter 2012; Imamura, Micha, et al. 2015.
13   Micha, Khatibzadeh, et al. 2015; Imamura, Micha, et al. 2015.
14   Micha, Khatibzadeh, et al. 2015.
15   La Vecchia and Majem 2015.
16   Lily Juo, 2015, "West Africans Have Some of the Healthiest Diets in the World," Quartz Africa, August 6, https://qz.com/473598/west-africans-have-some-of-the-healthiest-diets-in-the-world.
17   Vorster, Kruger, et al. 2011.
18   From a conversation with Mpho Tshukudu, quoted in Tshukudu and Trapido 2016.
19   Tshukudu and Trapido 2016; Haggblade, Duodu, et al. 2016.
20   Tshukudu and Trapido 2016.
21   Haggblade, Duodu, et al. 2016.
22   Popkin 2009.
23   Popkin 2009.
24   Popkin 2001; Popkin 2002; Popkin 2011; Popkin et al. 2012.
25   "About the Changing Global Diet," International Center for Tropical Agriculture (hereafter, CIAT), a CGIAR Research Center, accessed August 2018, https://ciat.cgiar.org/the-changing-global-diet/about/; Khoury, Bjorkman, et al. 2014.

26 "About the Changing Global Diet," CIAT, accessed August 2018, https:// ciat.cgiar.org/the-changing-global-diet/about/.

27 Khoury, Achinacoy, et al. 2016.

28 "About the Changing Global Diet," CIAT, accessed August 2018, https:// ciat.cgiar.org/the-changing-global-diet/about/.

29 Saladino 2017; Spector 2017.

30 Kammlade and Khoury 2017; Khoury 2017.

31 "Plants," Biodiversity Group, Food and Agriculture Organization of the United Nations, accessed July 2018, http://www.fao.org/biodiversity /components/plants/en/.

32 "Country Exploration," CIAT, accessed August 2018, https://ciat.cgiar .org/the-changing-global-diet/country-exploration/.

33 Mikkila, Vepsalainen, et al. 2015.

34 Mikkila, Vepsalainen, et al. 2015; Adair and Popkin 2012.

35 Khoury, Achicanoy, et al. 2016.

36 "The Mythical Banana Kingdom of Iceland," 2013, *Reykjavik Grape- vine*, December 2, accessed October 2018, https://grapevine.is/mag /articles/2013/12/02/the-mythical-banana-kingdom-of-iceland/; Kasper Friis, 2016, "A Banana Grows in Iceland," Atlas Obscura, March 8, http://www .atlasobscura.com/articles/bananas-in-iceland.

37 "Iceland's Bananas," Quite Interesting forum, accessed May 2017, http://old.qi.com/talk/viewtopic.php?t=33214&start=0&sid=7c40f741 2386dffcc75c72aa66bee5d6; "Mythical Banana Kingdom of Iceland," 2013.

38 "Mythical Banana Kingdom of Iceland," 2013.

39 "Banana: Statistical Compendium, 2015–16," Food and Agriculture Organization of the United Nations, accessed August 2018, http://www .fao.org/3/a-i7409e.pdf.

40 Dunn 2017.

41 Dunn 2017.

42 Lawrence 2004.

43 Rögnvaldardóttir 2002.

44 Rögnvaldardóttir 2002.

45 Lawrence 2004; Walvin 2018.

46 Roberts 2008.

47 Lawrence 2004; Roberts 2008.

48 Walvin 2008; Lang and Mason 2017.

49 Hawkes 2006.

50 "Country Exploration," CIAT, accessed July 2018, https://ciat.cgiar.org /the-changing-global-diet/country-exploration/.

51 Hawkes 2006.

52 Hawkes 2006.

53 Lopez and Jacobs 2018; Hawkes 2006; Eckhardt et al. 2006.

54 Doak, Adair, et al. 2005.

55 Bagni, Luis, et al. 2011.

56 Constance L. Hays and Donald G. McNeil Jr., 1998, "Putting Africa on Coke's Map: Pushing Soft Drinks on a Continent That Has Seen Hard, Hard Times," *New York Times*, May 26, https://www.nytimes .com/1998/05/26/business/putting-africa-coke-s-map-pushing-soft -drinks-continent-that-has-seen-hard-hard.html.

57 Jacobs and Richtel 2017.

58 "Door-to-Door Sales of Fortified Products," Nestlé website, accessed April 2018, https://www.nestle.com/csv/case-studies/allcasestudies/door-to-door salesoffortifiedproducts,brazil; Jacobs and Richtel 2017.

59 Popkin 2009; Kelly et al. 2010.

60 Popkin 2002; Wang, Leung, et al. 2014.

61 Kim, Moon, and Popkin 2000.

62 Lee, Popkin, and Kim 2002.

63 Lee, Popkin, and Kim 2002.

64 Lee, Popkin, and Kim 2002.

65 Kim, Moon, and Popkin 2000.

66 Popkin 2009; Kim, Moon, and Popkin 2000.

67 Lee, Duffey, and Popkin 2012.

68 Wiggins and Keats 2015.

69 Keats and Wiggins 2014.

70 Hahnemann 2016.

71 Keats and Wiggins 2014.

CHAPTER TWO: MISMATCH

1 Emily Rosen, "What Is Ancestral Eating?," Institute for the Psychology of Eating, accessed April 2018, http://psychologyofeating.com/ancestral -eating/.

2 Popkin, Adair, and Ng 2012.

3 Fresco 2015.

4   Yajnik 2018.

5   Yajnik 2018.

6   Yajnik, Fall, and Coyaji 2003.

7   Yajnik 2018.

8   Yajnik 2018.

9   Yajnik and Yudkin 2004.

10 Yajnik and Yudkin 2004.

11 "India's Missing Middle Class," 2018, *The Economist*, January 11, accessed April 2018, https://www.economist.com/news/briefing/21734382-multi national-businesses-relying-indian-consumers-face-disappointment-indias -missing-middle; Popkin, Adair, and Ng 2012.

12 "The New Face of Diabetes," *Vice,* produced by Elliot Kirschner, accessed October 2018, https://video.vice.com/en_us/video/the-new-face -of-diabetes/57fbfd04117c9766b44ad74b.

13 Nielsen and Popkin 2004.

14 Popkin 2009; Wolf, Bray, and Popkin 2008; Popkin and Hawkes 2016; DiMeglio and Mattes 2000.

15 Popkin 2009; Emiko Terazono and Neil Hume, 2016, "Are the Sweet Days Over for Orange Juice?" *Financial Times*, April 21, accessed October 2018, https://www.ft.com/content/c4bc7f92-0791-11e6-9b51-0fb5e65703ce.

16 Popkin 2009.

17 "Soft Drinks in Latin America: Keeping a Global Bright Spot Bright," 2014, Euromonitor International, January, accessed July 2018, http:// www.euromonitor.com/soft-drinks-in-latin-america-keeping-a-global -bright-spot-bright/report.

18 Wolf, Bray, and Popkin 2008.

19 Wolf, Bray, and Popkin 2008.

20 Wolf, Bray, and Popkin 2008; Mattes 2006.

21 Wolf, Bray, and Popkin 2008.

22 Mattes 2006.

23 Mattes 2006.

24 Victoria Richards, 2015, "Starbucks' New Frappuccinos Contain 'as Much Sugar as a Litre of Coke,'" *The Independent*, June 10, accessed June 2018, https://www.independent.co.uk/life-style/food-and-drink/news/starbucks -new-frappuccinos-contain-as-much-sugar-as-a-litre-of-coke-10310044 .html.

25 Richardson, Goodman, et al. 1961.

26   Cahnman 1968.

27   Brewis, Wutich, et al. 2011; Tomiyama 2014.

28   "Should We Officially Recognise Obesity as a Disease?" 2017.

29   Puhl and Heuer 2010.

30   Tomiyama 2014; Brewis et al. 2014.

31   Puhl and Heuer 2010.

32   Brewis 2014.

33   UConn Rudd Center for Food Policy & Obesity, accessed May 2018, www.uconnruddcenter.org.

34   Cahnman 1968.

CHAPTER THREE: EDIBLE ECONOMICS

1   De Vries, de Hoog, et al. 2016.

2   "The Changing Global Deit," CIAT, accessed August 2018, https://ciat .cgiar.org/the-changing-global-diet/.

3   Levy-Costa et al. 2005.

4   "The Changing Global Deit," CIAT, accessed August 2018, https://ciat .cgiar.org/the-changing-global-diet/.

5   Hawkes 2006.

6   Markley 1951.

7   Soskin 1988.

8   Hawkes 2006.

9   Simon Atkinson, 2017, "Why Are China Instant Noodle Sales Going Off the Boil?," BBC News, December 20, accessed July 2018, https://www .bbc.com/news/business-42390058.

10   Menzel and D'Aluiso 2005.

11   Hawkes 2004; Hawkes 2006.

12   Hawkes 2006.

13   Hawkes 2006.

14   Monteiro 2009; Monteiro, Cannon, et al. 2016.

15   Fiolet, Srour, et al. 2018.

16   Coudray 2017.

17   Fiolet, Srour, et al. 2018.

18   Monteiro, Moubarac, et al. 2013; Walvin 2018, 229.

19   Monteiro, Cannon, et al. 2016.

20   Sarah Boseley, 2018, "'Ultra-Processed' Products Now Half of All UK Family Food Purchases," *The Guardian*, February 2, accessed July 2018,

https://www.theguardian.com/science/2018/feb/02/ultra-processed
-products-now-half-of-all-uk-family-food-purchases.

21   Morley 2016.

22   Burnett 1983.

23   Burnett 1983; Anna-Louise Taylor, 2012, "Why Is Bread Britain's Most
Wasted Food?," BBC News, March 15, accessed November 2017, http://
www.bbc.co.uk/news/magazine-17353707.

24   "The Changing Global Diet," CIAT, accessed August 2018, https://ciat
.cgiar.org/the-changing-global-diet/.

25   Burnett 1983.

26   FAO.org, accessed November 2017.

27   Millstone and Lang 2008.

28   Natalie Lobel, 2017, "Bread's Not Bad for You. It's the Flour," The Daily
Meal, November 27, https://www.thedailymeal.com/cook/bread-s-not-bad
-you-it-s-flour.

29   Lang and Mason 2017.

30   Hansen 2013.

31   Clements and Chen 2009.

32   Clements and Chen 2009.

33   Clements and Chen 2009; USDA Economic Research Service, based on data
from Euromonitor; "Food Prices and Spending," United States Department
of Agriculture, accessed June 2018, https://www.ers.usda.gov/data-products
/ag-and-food-statistics-charting-the-essentials/food-prices-and-spending/.

34   USDA Economic Research Service, based on data from Euromonitor.

35   Hansen 2013; Wiggins and Keats 2015.

36   Hansen 2013, chapter 1.

37   Antoine Lewis, a.k.a. the Curly-Haired Cook, accessed August 2018,
https://antoinelewis.com/.

38   Fu, Ghandi, et al. 2012; Zaraska 2016.

39   Fu, Ghandi, et al. 2012; Zaraska 2016.

40   O'Brien 2013.

41   Hansen 2013; Lawler 2016.

42   Hansen 2013.

43   Rachel Hosie, 2017, "KFC's Double Down Burger: Is Bacon and Cheese
Sandwiched between Chicken as Good as It Sounds?," The Independent,
October 9, accessed November 2017, http://www.independent.co.uk/life
-style/food-and-drink/kfc-double-down-burger-uk-launch-chicken-bacon
-burger-taste-review-a7991121.html.

44   Lymbery and Oakeshott 2014, 167.

45   Lang and Mason 2017.

46   Lang and Mason 2017.

47   Jessica B. Harris, 2018, "Leah Chase: Queen of Creole Cuisine," *Garden & Gun*, August/September, accessed July 2018, https://gardenandgun.com /articles/leah-chase-queen-creole-cuisine/.

48   Tamar Haspel, 2017, "Junk Food Is Cheap and Healthful Food Is Expensive, but Don't Blame the Farm Bill," *Washington Post*, December 4, accessed August 2018, https://www.washingtonpost.com/lifestyle /food/im-a-fan-of-michael-pollan-but-on-one-food-policy-argument-hes -wrong/2017/12/04/c71881ca-d6cd-11e7-b62d-d9345ced896d_story .html?utm_term=.1a16a77fcd0a.

49   Wiggins and Keats 2015.

50   Wiggins and Keats 2015.

51   Cowen 2012.

52   Bloodworth 2018.

53   Powell and Bao 2009.

54   Wiggins and Keats 2015.

55   Wilson 2008.

CHAPTER FOUR: OUT OF TIME

1   Marmot and Syme 1976.

2   Quoted in Yano, Blackwelder, et al. 1979.

3   Jastran, Bisogni, et al. 2009.

4   Trentmann 2016, 443; "Average Annual Hours Actually Worked per Worker," Organization for Economic Co-Operation and Development, accessed April 2017, https://stats.oecd.org/Index.aspx?DataSetCode=ANHRS.

5   Trentmann 2016.

6   Email correspondence between Frank Trentmann and the author, April 2017.

7   Carroll 2013.

8   The countries were Belgium, Bulgaria, Estonia, Finland, France, Germany, Italy, Latvia, Lithuania, Norway, Poland, Slovenia, Spain, Sweden, and the United Kingdom.

9   Harmonised European Time Use Survey, "Areagraf—How Time Is Used During the Day," accessed November 2018, https://www.h6.scb.se/tus/tus /AreaGraphCID.html.

10 Brannen, O'Connell, and Mooney 2013.

11 Brannen, O'Connell, and Mooney 2013, 427.

12 Richard James, 2013, "Fat NHS Doctors Setting Bad Example in 'Poorly Developed' Obesity Services," *Metro*, January 1, http://metro.co .uk/2013/01/01/fat-nhs-doctors-setting-bad-example-in-poorly-developed -obesity-services-3333903/.

13 Bonnell, Huggins, et al. 2017.

14 Bonnell, Huggins, et al. 2017.

15 Townsend 2015.

16 Child 1832.

17 Robinson and Godbey 1997.

18 "India Tackles Food Waste Problem," 2014, video, BBC News, July 3, accessed May 2018, http://www.bbc.co.uk/news/av/business-28139586 /india-tackles-food-waste-problem.

19 Becker 1965.

20 Becker 1965.

21 Becker 1965.

22 Trentmann 2016.

23 Jabs and Devine 2006.

24 Becker 1965.

25 Pomiane 2008.

26 Kant and Graubard 2015.

27 Watt 2015.

28 C. Shivkumar, "Funds Drop State Securities from Portfolio," *Business Standard*, January 27, 2013, accessed February 2017, http://www .business-standard.com/article/specials/funds-drop-state-securities -from-portfolio-199101501061_1.html.

29 Zhai, Du, et al. 2014; Wang, Zhai, et al. 2012.

30 Wang, Horst, et al. 2018.

31 Popkin 2009.

32 Hawkes 2006.

33 Hawkes 2006.

34 Fisher, Wright, et al. 2015.

35 Hess and Slavin 2014.

36 Datamonitor 2015; Mintel 1985.

37 "As Snackification in Food Culture Becomes More Routine, Traditional Mealtimes Get Redefined," 2016, Hartman Group, February 16, accessed June 2017, http://www.hartman-group.com/hartbeat/638

/as-snackification-in-food-culture-becomes-more-routine-traditional
-mealtimes-get-redefined.

38 Choi, Choi, et al. 2008.

39 Tim Henderson, 2014, "More Americans Living Alone, Census Says," *Washington Post*, September 28, accessed December 2017, https://www .washingtonpost.com/politics/more-americans-living-alone-census -says/2014/09/28/67e1d02e-473a-11e4-b72e-d60a9229cc10_story .html?utm_term=.8d3a1f5216bf.

40 Hong 2016.

41 Basu 2016.

42 Manjoo 2017.

43 Hess 2017.

44 Hess 2017.

45 Johansen 2018.

46 Fulkerson, Larson, et al. 2014.

47 Oliver Burkeman, 2016, "Why Time Management Is Ruining Our Lives," *The Guardian*, December 22, accessed August 2018, https://www.theguardian .com/technology/2016/dec/22/why-time-management-is-ruining-our-lives.

CHAPTER FIVE: THE CHANGEABLE EATER

1 Dan Barber, 2009, "Tuscan Kale Chips," *Bon Appétit*, February, accessed April 2018, https://www.bonappetit.com/recipe/tuscan-kale-chips; Abend 2013; Mari Uyehara, 2017, "The 10th Anniversary of the Kale Salad as We Know It," *Taste*, October 24, accessed April 2018, https://www.tastecooking .com/10th-anniversary-kale-salad-know.

2 Kamp 2006.

3 "Get Ready for Some Serious Food Envy: The 20 Most Instagrammed Meals from around the World," 2016, *Daily Mail*, May 15.

4 Saffron Alexander, 2017, "Cloud Eggs: Instagram's Favourite New Food Fad," *Daily Telegraph*, May 8.

5 Jesse Szewczyk, 2017, "Rainbow Food Is Literally Garbage," BuzzFeed, April 7, accessed December 2017, https://www.buzzfeed.com/jesseszewczyk /its-official-hipsters-have-taken-rainbow-food-too-fucking?utm_term =.cndMEW5BD2#.ogQJ3Q2WMr.

6 Oliver 2016.

7 Harvey 2017.

8   Phil Lempert, 2016, "Food Trends vs. Food Fads," *Forbes*, June 16, accessed October 2017, https://www.forbes.com/sites/phillempert/2016/06/16/food-trends-vs-food-fads/#c0a619036550.

9   Sax 2014, introduction.

10 Van den Bos 2015.

11 Van den Bos 2015.

12 Van den Bos 2015; Mellentin 2018.

13 Mead 2013.

14 Lucy Rennick, 2018, "Why African Food Is the Next Big Thing," SBS, February 23, accessed July 2018, https://www.sbs.com.au/food/article/2018/02/23/why-african-food-next-big-thing.

15 "Now Comes Quinoa," 1954, *New York Times*, March 7.

16 Jacobsen 2011.

17 Tom Philpott, 2013, "Quinoa: Good, Evil, or Just Really Complicated?," *The Guardian*, January 25, accessed October 2017, https://www.theguardian.com/environment/2013/jan/25/quinoa-good-evil-complicated.

18 Hamilton 2014.

19 Lenny Flank, 2016, "Avocados and the Mexican Drug Cartels," Hidden History, July 8, accessed October 2017, https://lflank.wordpress.com/2016/07/08/avacados-and-the-mexican-drug-cartels/; "Mexico: Deforestation for Avocados Much Higher Than Thought," 2016, Associated Press, October 31, https://www.voanews.com/a/mexico-deforestation-avocados/3574039.html.

20 "Mexico's Avocado Army: How One City Stood Up to the Drug Cartels," 2017, *The Guardian*, May 18, accessed October 2017, https://www.theguardian.com/cities/2017/may/18/avocado-police-tancitaro-mexico-law-drug-cartels; Flank 2016.

21 Chris Elliott, 2014, "Elliott Review into the Integrity and Assurance of Food Supply Networks—Final Report: A National Food Crime Prevention Framework," July HM Government, accessed October 2017, https://www.gov.uk/government/uploads/system/uploads/attachment_data/file/350726/elliot-review-final-report-july2014.pdf; Whittle 2016.

22 Packer 2013.

23 Packer 2013.

24 "Global Coconut Water Market Forecast for Growth of Over 25%," 2017, FoodBev Media, February 7, accessed October 2017, https://www.foodbev.com/news/global-coconut-water-market-forecast-for-growth-of-over-25/.

25 David Derbyshire, 2017, "Coconut Oil: Are the Health Benefits a Big Fat Lie?," *The Guardian*, July 9, accessed October 2017, https://www .theguardian.com/lifeandstyle/2017/jul/09/coconut-oil-debunked-health -benefits-big-fat-lie-superfood-saturated-fats-lard.

26 Nestle 2018.

CHAPTER SIX: DINNER WITHOUT DUTY

1 Derek Thompson, 2013, "Cheap Eats: How America Spends Money on Food," *The Atlantic*, March 8, accessed October 2018, https://www.theat lantic.com/business/archive/2013/03/cheap-eats-how-america-spends -money-on-food/273811/.

2 Angela Monaghan, 2017, "Britons Spend More on Food and Leisure, Less on Booze, Smoking and Drugs," *The Guardian*, February 16, accessed August 2018, https://www.theguardian.com/business/2017/feb/16/britons -spending-more-on-food-and-leisure-than-booze-smoking-and-drugs.

3 Maumbe 2012; Nago, Lachat, et al. 2010.

4 Menzel and d'Aluisio 2005.

5 Popkin 2009.

6 Burnett 2004; Jacobs and Scholliers 2003.

7 Burnett 2004.

8 Lang and Millstone 2008.

9 "Wing Yip," Revolvy, accessed December 2017, https://www.revolvy.com /main/index.php?s=Wing%20Yip&item_type=topic; Burnett 2004.

10 Millstone and Lang 2008.

11 Warde and Martens 2009.

12 Lang and Mason 2017.

13 Guthrie 2002; Kant and Graubard 2004.

14 Orfanos et al. 2007.

15 "Veg Facts: A Briefing by the Food Foundation," 2016, The Food Foundation, November, accessed December 2017, http://foodfoundation.org.uk /wp-content/uploads/2016/11/FF-Veg-Doc-V5.pdf.

16 Krishnan, Coogan, et al. 2010.

17 Bahadoran, Mirmiran, et al. 2015; Nguyen and Powell 2014.

18 Krishnan, Coogan, et al. 2010.

19 Newman, Howlett, and Burton 2014; Currie, DellaVigna, et al. 2010.

20 Currie, DellaVigna, et al. 2010.

21 Hawkes 2006.

22 Simon Goodley, 2017, "Deliveroo Valuation Hits £1.5bn after Food Delivery Firm Riases New Funds," *The Guardian*, September 24, accessed December 2017, https://www.theguardian.com/business/2017/sep/24/deliveroo-valuation-hits-2bn-after-food-delivery-firm-raises-new-funds; Olson 2016.

23 "Consumer Spending on Pizza Delivery in the United States from 2004 to 2017 (in Billion US Dollars)," https://www.statista.com/statistics/259168/pizza-delivery-consumer-spending-in-the-us/; Dunn 2018.

24 Dunn 2018.

25 Khaleeli 2016.

26 Cuadra 2006; Food Stories, accessed December 2017, http://www.bl.uk/learning/resources/foodstories/index.html.

27 Tandoh 2018; Lang and Mason 2017.

28 Sole-Smith 2018.

29 Zhou, Du, et al. 2015; Tiffany C. Wright, "What Is the Profit Margin of a Supermarket?," azcentral, accessed August 2018, https://yourbusiness.azcentral.com/profit-margin-supermarket-17711.html.

30 Meades 2014.

31 Bowlby 2000.

32 Bowlby 2000.

33 Bowlby 2000.

34 Demmler, Ecker, et al. 2018.

35 "Eataly World and the Future of Food," Gastropodcast, October 10, 2017.

36 David 2010.

37 John Lanchester, 2018, "After the Fall," *London Review of Books* 40, no. 13, July 5, accessed August 2018, https://www.lrb.co.uk/v40/n13/john-lanchester/after-the-fall.

38 Cambridge Sustainable Food Hub, accessed December 2017, https://cambridgefoodhub.org/impacts/good-food-for-all/.

39 Biggs 2013.

40 "Foodbank Demand Soars across the UK," 2017, The Trussell Trust, November 7, accessed December 2017, https://www.trusselltrust.org/2017/11/07/foodbank-demand-soars-across-uk/.

41 Whitney Pipkin, 2016, "Why This Food Bank Is Turning Away Junk Food," Civil Eats, August 15, accessed August 2018, https://civileats.com/2016/08/15/why-this-food-bank-is-turning-away-junk-food/.

42   Hu 2016.
43   Hu 2016.

CHAPTER SEVEN: EATING BY THE RULES

1   Schwartz 2004.
2   Angela Palm, 2017 "Hierarchy of Needs," Longreads, Winter, accessed April 2018, https://longreads.com/2018/02/13/hierarchy-of-needs/.
3   Denise Lee Yohn, 2011, "Trader Joes, Where Less Is More," *Business Insider*, May 31, accessed April 2018, http://www.businessinsider.com/trader-joes-where-less-is-more-2011-5?IR=T; Andrea Felstead, "Day of the Discounters," *Financial Times*, December 10, 2014, accessed November 2018, https://www.ft.com/content/be5e8d52-7ec6-11e4-b83e-00144feabdc0.
4   Lindsay Whipp and Scheherezade Daneshkhu, "Big Business Identifies Appetite for Plant-Based Milk," *Financial Times*, July 15, 2016, accessed November 2018, https://www.ft.com/content/7df72c04-491a-11e6-8d68-72e9211e86ab.
5   "Is Celiac Disease on the Rise?," 2017, Beyond Celiac, June 1, accessed August 2018, https://www.beyondceliac.org/research-news/View-Research-News/1394/postid--81377/; Clifton, Carter, et al. 2015.
6   "Year-on-Year Growth of Top Volume Food Trend Dietary Restrictions Search Queries in the United States via Google as of February 2016," accessed June 2018, https://www.statista.com/statistics/612166/us-food-related-dietary-restrictions-searches/.
7   Laurie Budgar, 2011, "Veganism on the Rise among Health-Conscious Consumers," New Hope Network, June 8, accessed June 2018, http://www.newhope.com/food/veganism-rise-among-health-conscious-consumers; Dan Hancox, 2018, "The Unstoppable Rise of Veganism: How a Fringe Movement Went Mainstream," *The Guardian*, April 1, accessed September 2018, https://www.theguardian.com/lifeandstyle/2018/apr/01/vegans-are-coming-millennials-health-climate-change-animal-welfare.
8   Kateman 2017.
9   "Products," The Vegetarian Butcher, accessed November 2017, https://www.thevegetarianbutcher.com/products; Linette Lopez, 2016, "We Just Tried the 'Impossible Burger'—The Meatless Burger NYC Has Been Waiting For," *Business Insider*, July 27, accessed August 2018, http://uk.businessinsider.com/what-the-impossible-burger-tastes-like-2016-7.
10   McGregor 2017.

11 McGregor 2017.

12 Erikson 2004; Mintel data cited in email to author from Melissa Kvidahl.

13 De Crescenzo 2017.

14 Quoted in Kvidahl 2017.

15 Erikson 2004, 155.

16 Olivia Solon, 2017, "The Silicon Valley Execs Who Don't Eat for Days: 'It's Not Dieting, It's Biohacking,'" *The Guardian*, September 4, accessed September 2018, https://www.theguardian.com/lifeandstyle/2017/sep/04 /silicon-valley-ceo-fasting-trend-diet-is-it-safe.

17 Floris Wolswijk, 2016, "Soylent Eater Survey—The Results Are In!," Queal .com, June 20, accessed June 2018, https://queal.com/soylent-eater-survey -results/; Eliza Barclay, 2015, "Are Women Better Tasters Than Men?," NPR, August 31, accessed August 2017, https://www.npr.org/sections /thesalt/2015/08/31/427735692/are-women-better-tasters-than-men.

18 Widdecombe 2014.

CHAPTER EIGHT: THE RETURN TO COOKING

1 Perelman 2018.

2 Roberto A. Ferdman, 2015, "The Slow Death of the Home-Cooked Meal," *Washington Post*, March 5, accessed October 2018, https://www .washingtonpost.com/news/wonk/wp/2015/03/05/the-slow-death-of-the -home-cooked-meal/?utm_term=.b6ef41bff48e; Pollan 2013.

3 Pollan 2013, 3.

4 Conversation with author, November 1, 2017; Yoon 2017.

5 Conversation with author, November 1, 2017; Yoon 2017.

6 Pollan 2013.

7 Emmie Martin, 2017, "90 Percent of Americans Don't Like to Cook—and It's Costing Them Thousands Each Year," CNBC, September 27, accessed October 2018, https://www.cnbc.com/2017/09/27/how-much-americans -waste-on-dining-out.html.

8 Anita Singh, 2009, "Delia Effect Strikes Again," *The Telegraph*, December 3, accessed August 2018, https://www.telegraph.co.uk/culture /tvandradio/6709518/Delia-Effect-strikes-again.html; Marina O'Lough- lin, 2014, "Siam Smiles, Manchester—Restaurant Review," *The Guardian*, October 3, accessed August 2018, https://www.theguardian.com/life andstyle/2014/oct/03/siam-smiles-manchester-restaurant-review-marina -oloughlin.

9   Beth Kowitt, 2015, "The War on Big Food," *Fortune*, May 21, accessed
    October 2018, http://fortune.com/2015/05/21/the-war-on-big-food/;
    Caroline O'Donovan, 2016, "In Blue Apron's Chaotic Warehouses, Mak-
    ing Dinner Easy Is Hard Work," BuzzFeed News, October 2, accessed
    November 2017, https://www.buzzfeed.com/carolineodonovan/the-not
    -so-wholesome-reality-behind-the-making-of-your-meal?utm_term
    =.mjYOqRKla#.etQDGWN9g.

10 Caroline O'Donovan, 2016, "In Blue Apron's Chaotic Warehouses, Mak-
    ing Dinner Easy Is Hard Work," BuzzFeed News, October 2, accessed
    November 2017, https://www.buzzfeed.com/carolineodonovan/the-not
    -so-wholesome-reality-behind-the-making-of-your-meal?utm_term
    =.mjYOqRKla#.etQDGWN9g; Dan Orlando, 2017, "Meal Kits Will Help
    Amazon Infiltrate Traditional Grocery Market," July 17, accessed Octo-
    ber 2018, https://www.supermarketnews.com/online-retail/analyst-meal
    -kits-will-help-amazon-infiltrate-traditional-grocery-market.

11 Short 2006.

12 Rachel Roddy, 2016, "Rachel Roddy's Potato Gnocchi Recipe," *The Guard-
    ian*, January 12, accessed November 2017, https://www.theguardian.com
    /lifeandstyle/2016/jan/12/potato-gnocchi-recipe-rachel-roddy.

13 Laudan 2016.

14 Laudan 2016.

15 Matthew Holehouse, 2014, "Poor Going Hungry Because They Can't
    Cook, Says Tory Peer," *The Telegraph*, December 8, accessed August 2018,
    https://www.telegraph.co.uk/news/politics/11279839/Poor-going-hungry
    -because-they-cant-cook-says-Tory-peer.html.

16 Smith, Ng, et al. 2013; Adams and White 2015.

17 Kathleen Kerridge, 2017, "A Veg (or Five) Too Far: Why 10 Portions a
    Day Is Way Too Much to Ask," *The Guardian*, February 23, accessed
    June 2018, https://www.theguardian.com/commentisfree/2017/feb/23
    /austerity-britain-10-portions-fruit-and-veg.

18 Desai 2015.

19 Aribisala 2016.

20 Aribisala 2017.

21 Brigid Schulte, 2015, "What Gay Couples Get about Relationships that
    Straight Couples Often Don't," *Washington Post*, June 4, accessed May 2018,
    https://www.washingtonpost.com/news/wonk/wp/2015/06/04/what-gay

-couples-get-about-relationships-that-straight-couples-often-dont/?utm
_term=.24d60b53ddc1.

22 Ascione 2014.

CHAPTER NINE: CROSSING THE BRIDGE

1 "Global Hunger Index 2015 Fact Sheet," 2015, International Food Policy Research Institute, October 12, accessed August 2018, http://www
.ifpri.org/news-release/global-hunger-index-2015-fact-sheet; "World Hunger Again on the Rise, Driven by Conflict and Climate Change, New
UN Report Says," 2017, World Health Organization, September 15,
accessed April 2018, http://www.who.int/mediacentre/news/releases/2017
/world-hunger-report/en/.

2 Barber 2014.

3 Popkin 2009.

4 Felicity Lawrence, 2018, "Fat May Feel Like a Personal Issue—But Policy Is to Blame," *The Guardian*, May 11, accessed August 2018, https://
www.theguardian.com/commentisfree/2018/may/11/cutting-out
-chocolate-obesity-obesogenic-environment.

5 Michael M. Grynbaum, 2012, "New York Plans to Ban Sale of Big Sizes
of Sugary Drinks," *New York Times*, May 30, accessed August 2018,
https://www.nytimes.com/2012/05/31/nyregion/bloomberg-plans-a-ban
-on-large-sugared-drinks.html; Peter Roff, 2013, "Americans Don't Want
a Bloomburg Nannystate," *U.S. News*, March 12, accessed August 2018,
https://www.usnews.com/opinion/blogs/peter-roff/2013/03/12/bloomberg
-soda-ban-fail-a-victory-for-personal-freedom.

6 Cooper 2000; Ed Grover, 2016, "Researchers Call for Urgent Shift in
Food Research to Address World's 'Rising Nutrition Crisis,'" City University of London, November 30, accessed May 2018, https://www.city.ac.uk
/news/2016/november/researchers-call-for-urgent-shift-in-food-research
-to-address-worlds-rising-nutrition-crisis.

7 Caro, Ng, et al. 2017.

8 Andrew Jacobs, 2018, "In Sweeping War on Obesity, Chile Slays Tony the
Tiger," *New York Times*, February 7, accessed October 2018, https://www
.nytimes.com/2018/02/07/health/obesity-chile-sugar-regulations.html.

9 Jacobs 2018.

10   Hawkes, Smith, et al. 2015.

11   Bodzin 2014.

12   Deborah A. Cohen, 2018, "Fighting Obesity: Why Chile Should Continue Placing 'Stop Signs' on Unhealthy Foods," RAND, March 19, accessed September 2018, https://www.rand.org/blog/2018/03/fighting-obesity-why -chile-should-continue-placing-stop.html; Jacobs 2018.

13   Jacobs 2018.

14   Popkin and Hawkes 2016.

15   Colchero, Molina, et al. 2017.

16   Tina Rosenberg, 2015, "How One of the Most Obese Countries on Earth Took on the Soda Giants," *The Guardian*, November 3, accessed August 2018, https://www.theguardian.com/news/2015/nov/03/obese-soda-sugar -tax-mexico.

17   Behaviour Change by Design, accessed September 2018, https://www .behaviourchangebydesign.iph.cam.ac.uk.

18   Hollands, Shemilt, et al. 2013; Zupan, Evans, et al. 2017.

19   Bettina Elias Siegel, 2018, "Under Betti Wiggins, Houston ISD Signs $8 Million Contract for Domino's 'Smart Slice' Pizza," The Lunch Tray, August 2, accessed August 2018, https://www.thelunchtray.com /houston-isd-8-million-contract-for-dominos-smart-slice-pizza-betti -wiggins/.

20   "Flavour School: Making Sense of the Sapere Method," accessed August 2018, https://www.flavourschool.org.uk/.

21   "Time to Get Tough," Amsterdam.nl/zoblijvenwijgezond, December 2017.

22   "Healthy Weight Programme, Amsterdam: Urban Snapshot," NYC food policy newsletter, July 25, 2017.

23   Boseley 2017.

24   Boseley 2017.

25   "Time to Get Tough," Amsterdam.nl/zoblijvenwijgezond, December 2017.

26   Boseley 2017; "Amsterdam's Jump-In Programme," Obesity Action Scotland, accessed September 2018, http://www.obesityactionscotland.org /international-learning/amsterdam/amsterdams-jump-in-programme/.

27   "Time to Get Tough," Amsterdam.nl/zoblijvenwijgezond, December 2017; "Amsterdam Children Are Getting Healthier," City of Amsterdam, April 2017; "Review 2012–2017, Amsterdam Healthy Weight Programme."

28   Warren 1958.

29   "Veg Facts: A Briefing by the Food Foundation," 2016, Food Foundation, November, accessed June 2018, https://foodfoundation.org.uk/wp-content /uploads/2016/11/FF-Veg-Doc-V5.pdf.

30   "Veg Facts: A Briefing by the Food Foundation," 2016, Food Foundation, November, accessed June 2018, https://foodfoundation.org.uk/wp-content /uploads/2016/11/FF-Veg-Doc-V5.pdf.

31   "Veg Facts: A Briefing by the Food Foundation," 2016, Food Foundation, November, accessed September 2018, https://foodfoundation.org.uk/ wp-content/uploads/2016/11/FF-Veg-Doc-V5.pdf.

32   Sarika Bansal, 2012, "The Healthy Bodegas Intiative: Bringing Good Food to the Desert," *The Atlantic*, April 3, accessed September 2018, https://www.theatlantic.com/health/archive/2012/04/the-healthy-bodegas -initiative-bringing-good-food-to-the-desert/255061/.

33   Lloyd 2014.

34   Whitney 2017.

35   Whitney 2017.

36   Kludt and Geneen 2018.

37   Rao 2018.

38   Kludt and Geneen 2018.

39   Kirsten Aiken, 2018, "'White People Food' Is Creating an Unattainable Picture of Health," HuffPost, September 17, accessed August 2018, https:// www.huffingtonpost.co.uk/entry/white-people-food_us_5b75c270e 4b0df9b093dadbb.

40   @dmozaffarian, August 23, 2018.

41   "Food Is Medicine: Key Facts," Friedman School of Nutrition, Tufts University, accessed August 2018, https://nutrition.tufts.edu/sites/default/files /documents/FIM%20Infographic-Web.pdf.

42   Hawkes, Smith, et al. 2015.

EPILOGUE: NEW FOOD ON OLD PLATES

1   Bee Wilson, Jay Rayner, Tamal Ray, and Gizzi Erskine, 2016, "Our Gigantic Problem with Portions: Why Are We All Eating Too Much?," *The Guardian*, April 25, accessed May 2018, https://www .theguardian.com/lifeandstyle/2016/apr/25/problem-portions-eating -too-much-food-control-cutting-down.

2   Zupan, Evans, et al. 2017.

3   Sarah Boseley, 2018, "Ultra-Processed" Products Now Half of All UK
    Family Food Purchases," *The Guardian*, February 2, accessed May 2018,
    https://www.theguardian.com/science/2018/feb/02/ultra-processed-products
    -now-half-of-all-uk-family-food-purchases.

4   Shukler, Andono, et al. 2017.

5   Shukler, Andono, et al. 2017.

6   Spector 2015.

INDEX

Page numbers in italics indicate items in a figure.

BEE WILSON is a celebrated food writer, food historian, and author of five books, including *First Bite: How We Learn to Eat* and *Consider the Fork: A History of How We Cook and Eat.* She has been named BBC Radio's food writer of the year and is a three-time Guild of Food Writers food journalist of the year. She writes a monthly column on food in the *Wall Street Journal.* She lives in Cambridge, England.